D1758976

BOREHOLE ◇ FLOW MODELING

Wilson C. Chin

Gulf Publishing Company
Houston, London, Paris, Zurich, Tokyo

10 9 8 7 6 5 4 3 2

Gulf Publishing Company
Book Division
P.O. Box 2608, Houston, Texas 77252-2608

Library of Congress Cataloging-in-Publication Data

Chin, Wilson C.
    Borehole flow modeling in horizontal, deviated, and
vertical wells/Wilson C. Chin.
        p.   cm.
    Includes index.
    ISBN 0-88415-034-8
    1. Oil well drilling.   2. Wells—Fluid dynamics—
Mathematical models.   I. Title.
TN871.C4953 1991
622′.3381—dc20   91-25098
                                                          CIP

To Victoria, Jessica, and Jonathan

Contents

BOREHOLE ◇ Flow MODELING

Preface

Two seemingly unrelated oil service projects sparked my interest in annular flow during the early 1980s. In the first, I would determine cuttings size, shape and number density using acoustic backscatter data obtained from real-time, collar-mounted MWD sensors. And in the second, I would study mudcake erosion and dynamic filtration, insofar as they controlled drilling fluid invasion into oil-bearing sands. Once understood, displacement front information (extracted from resistivity measurements made in time) would provide porosity estimates directly.

Both projects, however, required good annular flow models which were lacking. The well known complications associated with eccentric geometries and nonlinear rheological effects were forbidding and therefore exciting from a research point of view. But my major impetus in tackling the general annular flow problem *correctly* came when I learned that the "low tech" problems brought about by hole cleaning posed the greatest operational hazards to a new technology known as horizontal drilling.

In 1986, I returned to the aerospace industry, and proceeded to develop the subject of annular flow simulation as rigorously as possible. The latest techniques in curvilinear grid generation and iterative relaxation methods were applied over several years to different aspects of the general problem, with the ultimate objective of bringing the methodology to the industry. Fortunately, I had worked with many of these mathematical methods as a Research Aerodynamicist at Boeing, incidentally, my first position as a new Ph.D. out of M.I.T.

The subject material covered in this book has formed the basis for a course in process engineering that I have taught at the University of Houston for the past five years. Thanks to my intellectually curious students, I am now able to present the subject of annular flow more coherently and clearly. The models derived in Chapters 2, 3 and 4, and applied to field problems in drilling and production in Chapter 5, are available in software form through Gulf Publishing Company in Houston. To its staff members, I express appreciation for the support and encouragement they have given. I would also like to thank my numerous colleagues with mud manufacturers, oil service and operating companies, who have generously shared their time, advice and data.

W. C. Chin
Houston, Texas

1
Overview of Annular Flow

This book describes three annular borehole flow models developed by the author. They were designed to handle the special problems that arise from drilling and producing deviated and horizontal wells, e.g., cuttings transport, stuck pipe, cementing and coiled tubing. In short, the models deal with (i) eccentric, nonrotating flow, (ii) concentric rotating flow, and (iii) recirculating heterogeneous flow. In this chapter, we will introduce the subject of borehole annular flow, briefly discuss the capabilities of the foregoing models, and describe operational problems which will benefit from detailed flow analysis.

BACKGROUND

The first model allows arbitrary eccentricity, assuming that the pipe (or casing) does not rotate. It solves the *complete* nonlinear viscous flow equations on "boundary conforming" grids, and does *not* invoke the "narrow annulus", "parallel plate" or "slot flow" assumptions commonly used. Holes with washouts, cuttings beds and square drill collars, for example, are easily simulated. The model is developed for Newtonian, power law, Bingham plastic and Herschel-Bulkley fluids.

The second model permits general pipe rotation, but it is restricted to concentric geometries. For Newtonian fluids, the results are shown to be *exact* solutions to the Navier-Stokes equations; both axial and azimuthal velocities satisfy no-slip conditions for all diameter ratios. For power law flows, a narrow annulus assumption is invoked which allows us to derive explicit closed form analytical solutions for all physical quantities. These formulas are easily programmed on pocket calculators. The results are checked against our

1

Newtonian formulas and shown to be consistent with these exact results in the "n=1" limit.

These models assume constant density, unidirectional axial flow where applied pressure gradients are exactly balanced by viscous stresses. These flows form the majority of observed fluid motions. But in deviated wells, especially where circulation has been temporarily interrupted, gravity segregation often causes weighting materials such as barite, fine cuttings and cement additives to fall out of suspension. The resulting density variations and inertial effects are primarily responsible for the strange "recirculating vortex flows" that have been experimentally observed from time to time. These isolated tornado-like clusters, completely fluid-dynamical in origin, are dangerous because they impede the mainstream flow; also, they entrain drilled cuttings and form stationary obstacles within the annulus.

Recirculating flows are stable packets of angular momentum that are wholly self-contained in stationary envelopes that sit in the midst of axial flows. The latter flows are, effectively, blocked. Within the envelopes are rotating fluid masses, some of which are roped off by closed streamlines; these highly three-dimensional flows are known to capture and trap solid particles and cuttings. Our third model describes these fascinating fluid motions and identifies the controlling nondimensional parameter. Computer simulations showing their generation and growth are given, and ways to avoid or eliminate their occurrence are suggested.

Although the mathematical models and numerical simulators have been available since 1987, publication was withheld pending application to field and laboratory examples. Often, the required data and empirical results were either unavailable or proprietary, contributing to delays in the evaluation of the work.

Experimental validation was crucial in establishing the credibility and accuracy of the computer modelling, especially because analytical solutions simply do not exist for the purposes of verification. Since numerical differencing methods, iteration and programming techniques invariably introduce additional assumptions which may be unphysical, consistency checks with empirical results were essential. Some of these extraneous effects include truncation error, mesh dependence and numerical viscosity.

Eventually, cuttings transport data, stuck pipe and other complementary information became available and the desired comparisons were undertaken after some initial delay. The first applications results were published in a series of articles, to be discussed, carried by Offshore Magazine beginning in 1990. An expanded "field oriented" treatment dealing with rigsite applications is offered in Chapter 5.

This book explains in detail the mathematical models and numerical algorithms used, provides calculated examples of "difficult" annular flows, and applies the computer models to problems related to hole cleaning, stuck pipe and cementing in deviated and horizontal wells. It is not essential to understand the details of the mathematics in order to appreciate the nuances of annular flow as uncovered by our calculations. In fact, the reader is encouraged to browse through the computed "snapshots" prior to any detailed study. Mathematics aside, the practical implications suggested by our examples will be understandable to most petroleum engineers.

The experienced researcher will have little trouble programming the flow models derived here. However, the algorithms are available to practitioners from Gulf Publishing Company in the form of source code and PC-executable software. The graphics modules described in Chapters 2, 3 and 4 are self-contained and require no additional software or hardware investment; and for eccentric geometries, annular shapes can be inputted as "sketches" drawn using a built-in ASCII file text editor. All numerical algorithms are written in standard Fortran so that they can be readily ported to different computer environments.

REVIEW OF PRIOR WORK

Annular flow analysis is important to drilling and production in deviated and horizontal wells. Different applications will be introduced shortly and covered in detail in Chapter 5. Despite their significance, few rigorous simulation models are available for research or field use.

There are several reasons for this dearth of analysis. First, the governing equations are nonlinear; this means that any useful solutions are necessarily numerical. Second, most practical annular geometries are complicated, making no-slip velocity boundary conditions difficult to enforce with any accuracy. Third, few computational algorithms are presently available for general rheologies that are stable, fast and robust. Consequently, researchers have chosen to study simpler although less realistic models whose mathematics are at least amenable to solution. These limitations have now been overcome, to some extent through technology transfer from related disciplines. Much of our work on eccentric flow, for example, represents an extension of aerospace industry research in simulating annular-like motions in jet engine ducts. Nonlinear equations are solved, for example, using fast iterative techniques developed by aerodynamicists for shear flows. And the work of Chapter 4 on heterogeneous flows draws, in part, upon the literature of dynamic meteorology and oceanography.

The model development summarized in this book is self-contained. The complete equations of motion for a fluid having a general stress tensor (Bird, Stewart and Lightfoot, 1960; Schlichting, 1968; Slattery, 1981; Streeter, 1961) are assumed. They are solved using physical boundary conditions relevant to petroleum applications (Gray and Darley, 1980; Moore, 1974; Whittaker, 1985;

Quigley and Sifferman, 1990; Govier and Aziz, 1977). The resulting formulations are solved using special relaxation methods and analytical techniques. An introduction to these methods may be found in Lapidus and Pinder (1982), Crochet et al. (1984) and Thompson et al. (1985). Let us review the existing literature on annular flow; no attempt is made to offer an exhaustive or comprehensive survey.

Modeling efforts may be classified into several increasingly sophisticated categories. The first contains simple models typified by References 13-16. The exact solution of Fredrickson and Bird (1958) falls into this category; it applies to concentric, nonrotating power law fluids. This well known numerical solution for an idealized rheology contrasts with another from the same class; namely that of Langlinais et al. (1983), an experimental study covering more complicated two-phase flows.

Eccentric annular flows are treated in the second category. To simplify the mathematics, most authors assume that the annulus is "almost concentric". This "parallel plate", "narrow annulus" or "slot flow" assumption is, for example, used in References 17-20. The results of these investigations are appealing because they provide a convenient analytical representation of the solution using elliptic integrals. However, their usefulness is severely limited because few eccentric annuli in deviated wells are "almost concentric".

Recently, Haciislamoglu and Langlinais (1990) importantly removed this slot flow restriction by reformulating the governing equations in bipolar coordinates. Just as circular polar coordinates imply simplifications to single well radial flow simulation, bipolar coordinates allow exact annular flow modeling of *circular* drillpipes and boreholes with arbitrary standoffs. The authors used an iterative finite difference method to model Bingham fluids, but they did not provide information on computing times and numerical stability or code portability. However, the methodology *cannot* be extended to handle boreholes with cuttings beds and washouts, or noncircular drillpipes and casings with stabilizers or centralizers.

In a second important paper, Haciislamoglu and Langlinais (1990) correctly pointed out that slot flow approaches typified by References 17-20 simulate radial shear only and neglect that component in the circumferential direction. These models, in other words, incorrectly use the equation for narrow concentric flow without accounting for the additional circumferential shear due to eccentricity.

The third category reverts to simpler concentric flows, but allows constant speed rotation. References 23-25 provide exact numerical solutions for power law fluids in concentric annuli. However, the solution techniques are cumbersome and not amenable for even research use. In Chapter 3, a simpler closed form "explicit" analytical solution is derived whose results agree with Savins and Wallick (1966) and Luo and Peden (1989a,b).

THE NEW ANNULAR FLOW MODELS

Eccentric, nonrotating annular flow. The need for fast, stable and accurate flow solvers for general eccentric annuli is central to drilling and production engineering. Because of mathematical difficulties - the nonlinearity of the governing equations and the complexity of most geometries - the problem is usually simplified by using unrealistic slot flow assumptions. Even then, the unwieldy elliptic integrals which result shed little physical insight into what remains of the problem. Moreover, the integrals require intensive computations, further decreasing their usefulness in field applications.

In Chapter 2, annular cross-sections containing eccentric circles are permitted. But importantly, the borehole contour may be modified "point by point" to simulate the effects of cuttings beds or wall deformations due to erosion and swelling. The pipe (or casing) contour may be likewise modified, for example, to model square drill collars or stabilizers and centralizers. Narrow annulus and slot flow assumptions are *not* invoked.

The analysis model handles Newtonian, power law, Bingham plastic and Herschel-Bulkley fluids. In all cases, the formulation satisfies no-slip velocity boundary conditions exactly at *all* solid surfaces. The model is derived from first principles using the general equations of continuum mechanics. The exact equations are rewritten in coordinates natural to the annular geometry under consideration. Then second-order accurate solutions for the axial velocity field, shear rate, shear stress, apparent viscosity, Stokes product and dissipation function are obtained. These solutions make use of recent developments in boundary conforming grid generation (Thompson et al., 1985) and relaxation methods (Lapidus and Pinder, 1982; Crochet et al., 1984).

Built-in graphics software allows computed results to be overlaid on the annular geometry itself, so that physical trends can be visually correlated with position. The unconditionally stable iteration process requires one minute per run on IBM PC-XTs equipped with math co-processors; the executable code occupies 100K RAM and does not require special graphics hardware or software. The Fortran code is portable and compatible with all machine environments. The computer program, written for generalists, uses "plain English" command menus only and requires the experience level of a novice petroleum engineer.

The only restriction is our assumption of a stationary nonrotating pipe. This is not overly severe in field applications, since the intended application of the model was anticipated in horizontal turbodrilled wells. Also, rotational effects will not be important when the tangential pipe speed is small compared to the average axial speed; estimates are easily obtained using the model of Chapter 3.

Concentric, rotating annular flow. Rotational effects are important when drilling at low flow rates, or when rotating the casing during cementing.

For eccentric annuli, all three velocity components will be nonlinearly coupled. Although numerical simulation is nevertheless possible, the simultaneous solution of three velocity equations requires computing resources not usually available at user locations. For this reason, the eccentric model of Chapter 2 is restricted to zero rotation.

Constant speed rotation, by contrast, can be treated quite generally for concentric annuli (References 23-25). However, the usual solution techniques do not lend themselves to simple use; the final equations are *implicit* and require iteration. Thus, in Chapter 3, the restriction to power law fluids in narrow annuli is made. Then a simple but powerful application of the Mean Value Theorem of differential calculus allows us to derive closed form solutions for the relatively complicated problem. Note that the problem still contains *four* coupled no-slip conditions, two for each of the axial and circumferential velocities.

The solutions are shown to be consistent with an exact solution of the Navier-Stokes equations for Newtonian flow, which does not bear any limiting geometric restrictions. Formulas for velocity, apparent viscosity, stress, deformation and dissipation function versus "r" are given. The solutions are *explicit* in that they require no iteration. A Fortran graphics algorithm is also described that conveniently plots and tabulates desired solutions, without requiring additional investment in graphical software and hardware.

Recirculating annular flow. The recirculating flows described earlier are interesting and fascinating in their own right. But they may be responsible for operational problems. In drilling, the presence of dynamically stable fluid-dynamic obstacles in the annular mainstream means that cuttings transport will be impaired. These stationary structures may affect bed build-up both upstream and downstream. In cementing applications, their presence in the mud or the cement slurry would suggest ineffective mud displacement. This implies poor zonal isolation and possibly the need for corrective squeeze cementing.

In any event, the transverse extent of any recirculation bubbles likely to be present is the ultimate solution sought. Thus, the model presented in Chapter 4 solves for streamline shapes and boundaries. Velocities, stresses and pressures can in principle be obtained from streamline patterns. The fractional blockage inferred from the presence of any closed streamlines provides a qualitative danger indicator for cuttings removal and cement displacement.

PROBLEMS IN DRILLING AND PRODUCTION

We introduce some practical applications for the annular flow simulators developed in Chapters 2, 3 and 4. These field applications for deviated and horizontal wells are listed and briefly discussed.

Cuttings Transport in Deviated Wells. The most important operational problem confronting drillers of deviated and horizontal wells is cuttings transport and bed formation. Many excellent experimental studies have been performed by industry and university groups, but the results are often confusing. For example, take eccentric inclined annular flows. Velocity, which plays an important role in vertical holes, has little value as a correlation parameter beyond a 30 deg deviation. Mean viscous stress turns out to be the parameter of significance; the right threshold value will erode cuttings beds formed at the bottom of the annulus. Extensive field data and computations support this view in Chapter 5.

What role do rotational viscometer measurements taken at the surface play in downhole applications? *None, because downhole shear rates - which change from case to case - are not known a priori.* Or consider pipe rotation. It turns out that concentric Newtonian flows, often used as the basis for convenient experiments, have no bearing to real world problems. In this singular limit, both the axial and azimuthal velocities decouple dynamically, and experimental observations cannot be extrapolated to other situations.

Spotting Fluids and Stuck Pipe. A related problem is the severe one dealing with stuck pipe. Typically, spotting fluids are used in combination with mechanical jarring motions to free immobile pipe. These impulsive transient flows can also be approximately analyzed, since the acceleration and pressure gradient terms in the momentum equation have like physical dimensions. What parameter governs spotting fluid effectiveness? How is that quantity related to lubricity? These questions are addressed in Chapter 5.

Coiled Tubing Return Flow. Sands and fines are often produced in flowing wells. They are removed by injecting non-Newtonian foams delivered by coiled tubing forced into the well. The debris is then transported up the return annulus inside the production tubing, a process not unlike the movement of drilled cuttings. Here the weight and small diameter of the metal tubing (typically, 1-in. to 2-in. O.D.) render the annular geometry highly eccentric. This problem is suited to the model of Chapter 2.

Cementing. Proper primary cementing creates the fluid seals needed to produce formation fluids properly. Improper procedures often lead to expensive and difficult squeeze cementing jobs. Mud that remains in the hole often does so because the cement velocity profile is hydrodynamically unstable, admitting viscous fingering and laminar flow breakdown. How are stable velocity profiles selected? How are dangerous "recirculation zones" that impede effective mud displacement eliminated? How does casing rotation alter the state of stress in the mud? These questions are addressed in Chapters 2-5.

Improved well planning. Traditional well planning involves mud pump selection and drilling fluid properties calculations. In vertical wells with concentric annuli, these are straightforward. But even simple questions require complicated answers for eccentric annular spaces, and particularly when they contain non-Newtonian fluids. "Can the pump operate through a range of mud weights and flow rates for a long horizontal well?" The dependence of flow rate on pressure gradient, of course, is nonlinear; and just as problematic is the apparent viscosity distribution, which depends on pressure gradient as well as annular geometry.

Borehole Stability. Borehole stability depends on several factors, principally mud chemistry and elastic states of stress. But annular flow can be important. For example, rapid velocities or surface stresses can erode borehole walls and promote washouts in unconsolidated sands. Drilling muds may also prove to be abrasive since they carry drilled cuttings that impinge into and damage the formation.

Wellbore Heat Generation. Temperature effects can be important in drilling. While heat generation due to internal friction is generally small, overall temperature increases within a closed system may be significant for large circulation times. These may affect the thermal stability and thinning of oil base muds.

Heat generation may be important to temperature log interpretation. To correctly extrapolate formation temperature from measurements obtained while drilling, it is necessary to correct for temperature effects due to total circulation time and internal friction due to rheology.

Summary. This book significantly expands upon the author's previously published articles in Offshore Magazine (Chin, 1990a,b,c; Chin, 1991). The reader may refer to those publications for a quick synopsis of the present work.

PHILOSOPHY ON NUMERICAL METHODS

Reservoir engineers and structural dynamicists, for example, routinely use advanced finite difference and finite element methods. But drillers have traditionally relied upon simpler handbook formulas and tables that are convenient at the rigsite. Simulation methods are powerful, to be sure, but they also have their limitations. This section explains the pitfalls and the philosophy one must adopt in order to bring state-of-the-art techniques to the field.

Numerical methods do not always yield exact answers. But more often than not, they produce excellent *trend information* that is useful in practical application. For our purposes, consider the steady, concentric annular flow of a Newtonian fluid (Bird et al., 1960). The governing equations are

$$d^2u(r)/dr^2 + r^{-1} du/dr = 1/\mu \; dp/dz \tag{1-1}$$

$$u(R_i) = u(R_o) = 0 \tag{1-2}$$

In Equations 1-1 and 1-2, $u(r)$ is the annular speed satisfying no-slip conditions at the inner and outer radii, R_i and R_o. The viscosity μ and the applied pressure gradient dp/dz are known constants. The exact solution for this linear formulation is derived in Chapter 3.

Here we will examine the consequences of a numerical solution. A second-order accurate scheme is easily derived by "central differencing" Equation 1-1 as follows,

$$(u_{j-1} - 2u_j + u_{j+1}) / (\delta r)^2 + (u_{j+1} - u_{j-1})/2r_j \delta r = 1/\mu \; dp/dz \tag{1-3}$$

where u_j refers to the value of $u(r)$ at the jth radial node at the r_j th location, j being an ordering index. Equation 1-3 can be evaluated at any number of interior nodes for the mesh length δr. The resulting difference equations, when augmented by the no-slip condition

$$u_1 = u_{jmax} = 0 \tag{1-4}$$

obtained using Equation 1-2, form a tridiagonal system of j_{max} unknowns that lends itself to simple solution. The resulting u_j's can be post-processed to determine the net volume flow rate.

For our first run, we assumed $R_i = 4$ inch, $R_o = 5$ inch, $dp/dz = -0.0005$ psi/in and $\mu = 2$ cp. Computed flow rates as functions of the number of meshes used are given in Table 1-1. Note how the "100 mesh" solution is almost exact; but the "10 mesh" solution for flow rate, which is ten times faster to compute, is satisfactory for engineering purposes. Now let us double the viscosity μ and recompute the solution. The gpms so obtained decrease exactly by a factor of two, and the dependence on viscosity is certainly brought out very clearly.

However, the *trend information* relating changes in gpm to those in μ are accurately captured even for coarse meshes. So, sometimes fine meshes are unnecessary. Similar comments apply to the pressure gradient dp/dz.

Table 1-1
Volume Flow Rate versus Mesh Number

# Meshes	GPM	% Error
2	783	25
3	929	11
4	980	6
5	1003	4
10	1035	1
20	1042	0
30	1044	0
100	1045	0

It is clear that the exact value of u(r) is mesh dependent; the finer the mesh, the better the answer. In some applications, it may be essential to find, through trial and error, a mesh distribution that leads to the exact solution or that is consistent with real data in some engineering sense. From that point on, "what if" analyses may be performed accurately with confidence. This rationale is used in reservoir engineering, where history matching with production data plays a crucial role in estimating reserves.

For other applications, the exact numbers may not be as important as qualitative trends caused by changing different physical parameters. For example, how does hole eccentricity affect volume flow rate for a prescribed pressure gradient? For a given annular geometry, how does a decrease in the power law exponent affect velocity profile curvature?

In structural engineering, it is well known that *uncalibrated* finite element analyses can accurately pinpoint *where* cracks are likely to form even though the computed stresses may not be correct. For such qualitative objectives, the results of numerical solution may be accepted "as is" provided the calculated numbers are not literally interpreted. Agreement with exact solutions, of course, is important; but often it is the very lack of such analytical solutions itself that motivates numerical alternatives. Thus, while consistency with exact solutions is desirable, in practice it is through the use of *comparative solutions* that computational methods offer their greatest value.

For annular flows, this philosophy is appropriate because there are no analytical solutions or detailed laboratory measurements with which to establish standards for comparison. We should be satisfied as long as our solutions agree roughly with field data; our real objective, remember, aims at establishing *trends* with respect to *changes* in parameters like fluid rheology, flow rate and hole eccentricity.

We will show through extensive computations and correlation with empirical data that the models developed in Chapters 2, 3 and 4 are correct and useful in this engineering sense. The ultimate acid test lies in field applications, and these are addressed in Chapter 5.

We emphasize that the eccentric flow of Chapter 2, the main thrust of this book, is by no means as simple as the above example might suggest. In Equation 1-1, the unknown speed u(r) depends on a single variable "r" only. In Chapter 2, the velocity depends on two cross-sectional coordinates x and y; this leads to a partial differential equation. The "two-point" boundary conditions in Equation 1-2 are therefore replaced by no-slip velocity conditions enforced along two general arbitrary closed curves representing the borehole and drillpipe contours. To implement these no-slip conditions accurately, "boundary conforming meshes" must be used that provide high resolution in tight spaces. To be computationally efficient, these meshes must be *variable* with respect to all coordinate directions.

The difference equations solved on such host meshes must be solved *iteratively*; for unlike Equations 1-3 and 1-4, which apply to Newtonian flows with constant viscosities, the power law, Bingham plastic and Herschel-Bulkley fluids considered in this book satisfy nonlinear equations with problem-dependent apparent viscosities. The algorithms must be *fast, stable and robust*; they must produce solutions without straining computing resources. Finally, the computed solutions must be physically correct; this is perhaps the final arbiter that challenges all numerical simulations.

REFERENCES

---

1. Bird, R.B., Stewart, W.E., and Lightfoot, E.N., *Transport Phenomena*, New York: John Wiley and Sons, 1960.
2. Schlichting, H., *Boundary Layer Theory*, New York: McGraw-Hill, 1968.
3. Slattery, J.C., *Momentum, Energy, and Mass Transfer in Continua*, New York: Robert E. Krieger Publishing Company, 1981.
4. Streeter, V.L., *Handbook of Fluid Dynamics*, New York: McGraw-Hill, 1961.
5. Gray, G.R., and Darley, H.C.H., *Composition and Properties of Oil Well Drilling Fluids*, Houston: Gulf Publishing Company, 1980.
6. Moore, P. L., *Drilling Practices Manual*, Tulsa: PennWell Books, 1974.

7.   Whittaker, A., *Theory and Application of Drilling Fluid Hydraulics*, Boston: IHRDC Press, 1985.

8.   Quigley, M.S., and Sifferman, T.R., "Unit Provides Dynamic Evaluation of Drilling Fluid Properties," *World Oil*, January 1990, pp. 43-48.

9.   Govier, G.W. and Aziz, K., *The Flow of Complex Mixtures in Pipes*, New York: Robert E. Krieger Publishing Company, 1977.

10. Lapidus, L., and Pinder, G., *Numerical Solution of Partial Differential Equations in Science and Engineering*, New York: John Wiley and Sons, 1982.

11. Crochet, M.J., Davies, A.R., and Walters, K., *Numerical Simulation of Non-Newtonian Flow*, Amsterdam: Elsevier Science Publishers B.V., 1984.

12. Thompson, J.F., Warsi, Z.U.A., and Mastin, C.W., *Numerical Grid Generation*, New York: Elsevier Science Publishing, 1985.

13. Fredrickson, A.G., and Bird, R.B., "Non-Newtonian Flow in Annuli," *Ind. Eng. Chem.*, 1958, Vol. 50, p. 347.

14. Savins, J.G., "Generalized Newtonian (Pseudoplastic) Flow in Stationary Pipes and Annuli," *Petroleum Transactions*, AIME, Vol. 213, 1958, pp. 325-332.

15. Zamora, M., and Lord, D.L., "Practical Analysis of Drilling Mud Flow in Pipes and Annuli," *SPE Paper 4976, 49th Annual Technical Conference and Exhibition of the Society of Petroleum Engineers*, Houston, October 6-9, 1974.

16. Langlinais, J.P., Bourgoyne, A.T., and Holden, W.R., "Frictional Pressure Losses for the Flow of Drilling Mud and Mud/Gas Mixtures," *SPE Paper 11993, 58th Annual Technical Conference and Exhibition of the Society of Petroleum Engineers*, San Francisco, October 5-8, 1983.

17. Vaughn, R.D., "Axial Laminar Flow of Non-Newtonian Fluids in Narrow Eccentric Annuli," *Society of Petroleum Engineers Journal*, December 1965, pp. 277-280.

18. Iyoho, A.W., and Azar, J.J., "An Accurate Slot-Flow Model for Non-Newtonian Fluid Flow Through Eccentric Annuli," *Society of Petroleum Engineers Journal*, October 1981, pp. 565-572.

19. Luo, Y., and Peden, J.M., "Flow of Drilling Fluids Through Eccentric Annuli," *Paper 16692, 1987 SPE Annual Technical Conference and Exhibition, Dallas*, September 27-3, 1987.

20. Uner, D., Ozgen, C., and Tosun, I., "Flow of a Power Law Fluid in an Eccentric Annulus," *SPE Drilling Engineering*, September 1989, pp. 269 - 272.

21. Haciislamoglu, M., and Langlinais, J., "Non-Newtonian Fluid Flow in Eccentric Annuli," *1990 ASME Energy Resources Conference and Exhibition*, New Orleans, January 14-18, 1990.

22. Haciislamoglu, M., and Langlinais, J., "Discussion of Flow of a Power-Law Fluid in an Eccentric Annulus," *SPE Drilling Engineering*, March 1990, p. 95.

23. Savins, J.G., and Wallick, G.C., "Viscosity Profiles, Discharge Rates, Pressures, and Torques for a Rheologically Complex Fluid in a Helical Flow," *A.I.Ch.E. Journal*, Vol. 12, No. 2, March 1966, pp. 357-363.

24. Luo, Y., and Peden, J.M., "Laminar Annular Helical Flow of Power Law Fluids, Part I: Various Profiles and Axial Flow Rates," *SPE Paper 20304,* December 1989.

25. Luo, Y., and Peden, J.M., "Reduction of Annular Friction Pressure Drop Caused by Drillpipe Rotation," *SPE Paper 20305*, December 1989.

26. Chin, W.C., "Advances in Annular Borehole Flow Modeling," *Offshore Magazine*, February 1990, pp. 31-37.

27. Chin, W.C., "Exact Cuttings Transport Correlations Developed for High Angle Wells," *Offshore Magazine*, May 1990, pp. 67-70.

28. Chin, W.C., "Annular Flow Model Explains Conoco's Borehole Cleaning Success," *Offshore Magazine*, October 1990, pp. 41-42.

29. Chin, W.C., "Model Offers Insight into Spotting Fluid Performance," *Offshore Magazine*, February 1991, pp. 32-33.

2
Eccentric, Nonrotating Annular Flow

Numerical solutions for the nonlinear, two-dimensional axial velocity field, and its corresponding stress and shear rate distributions, are obtained for the eccentric annular flow in an inclined borehole. The homogeneous fluid is assumed to be flowing unidirectionally in a wellbore containing a nonrotating drillstring. The unconditionally stable algorithm used draws upon finite difference relaxation methods (Lapidus and Pinder, 1982; and Crochet, Davies and Walters, 1984), and recent advances in differential geometry and boundary conforming grid generation (Thompson, Warsi and Mastin, 1985).

Slot flow, narrow annulus and parallel plate assumptions are not invoked. The cross-section may contain conventional concentric or nonconcentric circular drillpipes and boreholes. But importantly, the hole and pipe contours may be arbitrarily modified "point by point" to simulate the effects of square drill collars, centralizers, stabilizers, thick cuttings beds, and general wall deformations due to swelling and erosion.

The overall formulation, which applies to general rheologies, is specialized to Newtonian flows, power law fluids, Bingham plastics and Herschel-Bulkley flows. In all instances, no-slip velocity boundary conditions are satisfied exactly at all solid surfaces. Detailed spatial solutions and cross-sectional plots for local annular velocity, apparent viscosity, two components each of viscous stress and shear rate, Stokes product and heat generation due to fluid friction, are presented for a large number of complicated annular geometries. Net volume flow rates are also given.

14

Calculated results are displayed using a special "character-based" text graphics program that overlays computed quantities on the annular cross-section itself, thus facilitating the physical interpretation and visual correlation of numerical quantities with annular position. For most annular geometries of practical interest, mesh generation requires approximately one minute of computing time on IBM PC-XTs with math co-processors. Once the host mesh is available, any number of "what if" scenarios for differing rheologies or net flow rates can be efficiently evaluated, these simulations again requiring one minute.

In this chapter, the basic ideas are derived from first principles and explained mathematically. However, the reader who is more interested in practical applications may, without loss of continuity, proceed directly to "Example Calculations" and to Chapter 5.

THEORY AND MATHEMATICAL FORMULATION

The equations governing general fluid motions in three spatial dimensions are available from many excellent textbooks (Bird, Stewart and Lightfoot, 1960; Schlichting, 1968; Slattery, 1981; and Streeter, 1961). We will cite these equations without proof. Let u, v and w denote Eulerian fluid velocities, and F_z, F_y and F_x denote body forces, in the z, y and x directions respectively, where (z,y,x) are Cartesian coordinates.

Also, let ρ be the constant fluid density and p be the pressure; and denote by S_{zz}, S_{yy}, S_{xx}, S_{zy}, S_{yz}, S_{xz}, S_{zx}, S_{yx} and S_{xy} the nine elements of the general extra stress tensor $\underline{\underline{S}}$. If t is time and ∂'s represent partial derivatives, the complete equations of motion obtained from Newton's law and from mass conservation are, without approximation,

Momentum equation in z:

$$\rho \ (\partial u/\partial t + u \ \partial u/\partial z + v \ \partial u/\partial y + w \ \partial u/\partial x) =$$
$$= F_z - \partial p/\partial z + \partial S_{zz}/\partial z + \partial S_{zy}/\partial y + \partial S_{zx}/\partial x \qquad (2\text{-}1)$$

Momentum equation in y:

$$\rho \ (\partial v/\partial t + u \ \partial v/\partial z + v \ \partial v/\partial y + w \ \partial v/\partial x) =$$
$$= F_y - \partial p/\partial y + \partial S_{yz}/\partial z + \partial S_{yy}/\partial y + \partial S_{yx}/\partial x \qquad (2\text{-}2)$$

Momentum equation in x:

$$\rho \; (\partial w/\partial t + u \; \partial w/\partial z + v \; \partial w/\partial y + w \; \partial w/\partial x) =$$
$$= F_x - \partial p/\partial x + \partial S_{xz}/\partial z + \partial S_{xy}/\partial y + \partial S_{xx}/\partial x \qquad (2\text{-}3)$$

Mass continuity equation:

$$\partial u/\partial z + \partial v/\partial y + \partial w/\partial x = 0 \qquad (2\text{-}4)$$

These equations apply to all Newtonian and non-Newtonian fluids. In continuum mechanics, the most common class of empirical models for incompressible fluids assumes that $\underline{\underline{S}}$ can be related to the rate of deformation tensor $\underline{\underline{D}}$ by a relationship of the form

$$\underline{\underline{S}} = 2 \, N(\Gamma) \, \underline{\underline{D}} \qquad (2\text{-}5)$$

where the elements of $\underline{\underline{D}}$ are

$$D_{zz} = \partial u/\partial z \qquad (2\text{-}6)$$

$$D_{yy} = \partial v/\partial y \qquad (2\text{-}7)$$

$$D_{xx} = \partial w/\partial x \qquad (2\text{-}8)$$

$$D_{zy} = D_{yz} = (\partial u/\partial y + \partial v/\partial z)/2 \qquad (2\text{-}9)$$

$$D_{zx} = D_{xz} = (\partial u/\partial x + \partial w/\partial z)/2 \qquad (2\text{-}10)$$

$$D_{yx} = D_{xy} = (\partial v/\partial x + \partial w/\partial y)/2 \qquad (2\text{-}11)$$

assuming isotropic flow. In Equation 2-5, $N(\Gamma)$ is the well known "apparent viscosity function" satisfying

$$N(\Gamma) > 0 \qquad (2\text{-}12)$$

where $\Gamma(z,y,x)$ is the scalar functional of u, v and w defined by the tensor operation

$$\Gamma = \{ \, 2 \text{ trace } (\underline{\underline{D}}o\underline{\underline{D}}) \, \}^{1/2} \qquad (2\text{-}13)$$

Unlike the constant laminar viscosity in Newtonian flow, the apparent viscosity in general depends on the details of the particular problem under consideration. These details include the rheological model used, the exact annular geometry occupied by the fluid, and the applied pressure gradient or the net volume flow rate. Also, the apparent viscosity varies with the position (z,y,x) in the annular domain.

These considerations are still very general. To fix ideas, let us examine one important and practical simplification. The Ostwald-de Waele model for two-parameter "power law" fluids assumes that the apparent viscosity satisfies

$$N(\Gamma) = k \, \Gamma^{n-1} \tag{2-14a}$$

where the "consistency factor" k and the "fluid exponent" n are constants. Such power law fluids are "pseudoplastic" when $0 < n < 1$, Newtonian when $n = 1$, and "dilatant" when $n > 1$. Most drilling fluids are pseudoplastic. In the limit $(n=1, k=\mu)$, Equation 2-14a reduces to the classical Newtonian model with $N(\Gamma) = \mu$, where μ is the constant laminar viscosity; in this limit, stress is directly proportional to the rate of strain.

Power law and Newtonian fluids respond instantaneously to applied pressure and stress. But if the fluid behaves as a rigid solid until the net applied stresses have exceeded some known critical yield value, say S_{yield}, then Equation 2-14a can be generalized by writing

$$N(\Gamma) = k\Gamma^{n-1} + S_{yield}/\Gamma \text{ if } \{1/2 \text{ trace } (\underline{\underline{S}} o \underline{\underline{S}})\}^{1/2} > S_{yield}$$

$$\underline{\underline{D}} = 0 \text{ if } \{1/2 \text{ trace } (\underline{\underline{S}} o \underline{\underline{S}})\}^{1/2} < S_{yield} \tag{2-14b}$$

In this form, Equation 2-14b rigorously describes the general Herschel-Bulkley fluid. When the additional "Newtonian-like" limit $(n=1, k=\mu)$ is taken, the first formula in Equation 2-14b specializes to

$$N(\Gamma) = \mu + S_{yield}/\Gamma \text{ if } \{1/2 \text{ trace } (\underline{\underline{S}} o \underline{\underline{S}})\}^{1/2} > S_{yield} \tag{2-14c}$$

This is the well known Bingham plastic model, where μ represents the "plastic viscosity". Annular flows containing fluids with nonzero yield stresses are more difficult to analyze, both mathematically and numerically, than those marked by zero yield. This is so because there may coexist "dead", "plug" and "shear" flow regimes with internal boundaries that must be determined as part of the solution.

We will restrict our discussion first to Newtonian and power law flows, that is to fluids without yield stresses. For isotropic flows whose velocities do not depend on the axial coordinate z, and which further satisfy $v = w = 0$, the functional Γ in Equation 2-14a takes the form

$$\Gamma = [(\partial u/\partial y)^2 + (\partial u/\partial x)^2]^{1/2} \tag{2-15}$$

so that Equation 2-14a becomes

$$N(\Gamma) = k \, [\, (\partial u/\partial y)^2 + (\partial u/\partial x)^2 \,]^{(n-1)/2} \tag{2-16}$$

This expression for the apparent viscosity reduces to the conventional "$N(\Gamma) = k \, (\partial u/\partial y)^{(n-1)}$" formula for the simpler one-dimensional, parallel plate flows considered in the literature.

When both independent variables y and x for the cross-section are present, as is the case for eccentric annular flow, significant mathematical difficulty arises as discussed in Chapter 1. For one, the original ordinary differential equation (ODE) for annular velocity becomes a partial differential equation (PDE). And whereas the ODE requires boundary conditions at two points, the PDE requires no-slip boundary conditions along two arbitrary closed curves in doubly-connected space. The nonlinearity of the governing PDE and the irregular annular geometry only compound these difficulties.

The hole configuration considered is shown in Figure 2-1. A drillpipe (or casing) and borehole combination is inclined at an angle α relative to the ground, with $\alpha = 0^o$ for horizontal wells and $\alpha = 90^o$ for vertical ones. Here "z" denotes any point within the annular fluid; Section "AA" is a cut taken normal to the local z axis. Figure 2-2 resolves the vertical body force due to gravity at "z" into components parallel and perpendicular to the borehole axis. Figure 2-3 provides a detailed picture of the annular cross-section at Section "AA". Physical assumptions about the drillstring and borehole flow in these coordinates are discussed next.

We will specialize the foregoing equations to downhole annular flows. In Figures 2-1, 2-2 and 2-3, we have aligned z, which increases downward, with the axis of the borehole. The plane of the variables (y,x) is perpendicular to the z-axis, and (z,y,x) are mutually orthogonal Cartesian coordinates. Thus the body forces due to the gravitational acceleration g can be resolved into components

$$F_z = \rho \, g \sin \alpha \tag{2-17}$$

$$F_x = - \rho \, g \cos \alpha \tag{2-18}$$

$$F_y = 0 \tag{2-19}$$

If we now assume that the drillpipe does not rotate, the resulting fluid can only flow in a direction parallel to the borehole axis. This requires that the velocities v and w vanish. Therefore,

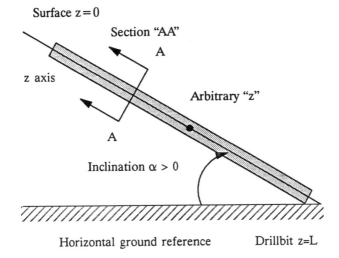

Surface z=0

Section "AA"

z axis

Arbitrary "z"

Inclination α > 0

Horizontal ground reference   Drillbit z=L

Figure 2-1. Borehole Configuration.

$$v = w = 0 \tag{2-20}$$

We have restricted our analysis to constant density flows, that is

$$\partial\rho/\partial t = 0 \tag{2-21}$$

Equations 2-4, 2-20 and 2-21 together imply that the axial velocity u = u(y,x,t) does not depend on z. And if we further confine ourselves to steady flow, that is to flows driven by axial pressure gradients that do not vary in time, we find that

$$u = u(y,x) \tag{2-22}$$

depends at most on two independent variables, namely the cross-sectional coordinates y and x.

In the case of a concentric drillpipe and borehole, it is more convenient to collapse y and x into a radial coordinate "r". This is accomplished by using the definition $r = (x^2 + y^2)^{1/2}$. For general eccentric flows, the lack of similar algebraic transformations drives the use of grid generation methods. Next substitution of Equations 2-20 and 2-22 into Equations 2-1, 2-2 and 2-3 leads to

$$0 = \rho g \sin \alpha \; - \partial p/\partial z \; + \partial S_{zy}/\partial y \; + \partial S_{zx}/\partial x \tag{2-23}$$

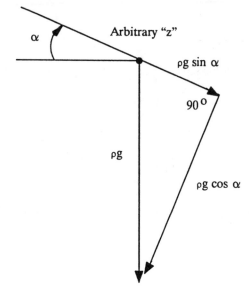

Figure 2-2. Gravity Vector Components.

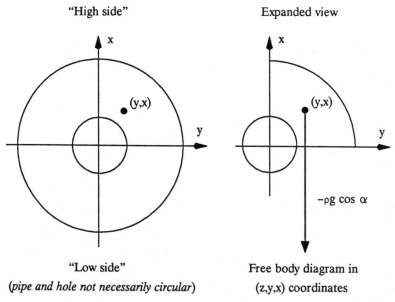

Figure 2-3. Gravity Vector Components.

$0 = - \partial p/\partial y$ (2-24)

$0 = -\rho g \cos \alpha - \partial p/\partial x$ (2-25)

If we introduce, without loss of generality, the pressure separation of variables

$P = P(z,x) = p - z\rho g \sin \alpha + x\rho g \cos \alpha$ (2-26)

we can replace Equations 2-23, 2-24 and 2-25 by the single equation

$\partial S_{zy}/\partial y + \partial S_{zx}/\partial x = \partial P/\partial z = \text{constant}$ (2-27)

where the constant pressure gradient $\partial P/\partial z$ is prescribed.

Now recall the definitions of the deformation tensor elements given in Equations 2-6 to 2-11 and the fact that $\underline{\underline{S}} = 2N\underline{\underline{D}}$ to rewrite Equation 2-27 in the form

$\partial(N \partial u/\partial y)/\partial y + \partial(N \partial u/\partial x)/\partial x = \partial P/\partial z$ (2-28)

Here the function $N(\Gamma)$ is, without approximation, given by the nonlinear equation

$N(\Gamma) = k [(\partial u/\partial y)^2 + (\partial u/\partial x)^2]^{(n-1)/2}$ (2-29)

Equations 2-28 and 2-29 comprise the entire system to be solved along with no-slip velocity boundary conditions at drillpipe and borehole surfaces.

It is important, for the purposes of numerical analysis, to recognize how Equation 2-28 can be written as a nonlinear Poisson equation. After some manipulation, we obtain

$$\partial^2 u/\partial y^2 + \partial^2 u/\partial x^2 = [\partial P/\partial z + (1-n)N(\Gamma)(u_y^2 u_{yy}$$
$$+2u_y u_x u_{yx} + u_x^2 u_{xx})/(u_y^2 + u_x^2)] / N(\Gamma)$$ (2-30)

In this form, conventional solution techniques for elliptic equations can be employed. These might include iterative techniques as well as direct inversion methods. The nonlinear terms in the square brackets, for example, can be evaluated using latest values in a successive approximations scheme.

Also various algebraic simplifications are possible. For some values of n, particularly those near unity, these nonlinear terms may represent negligible higher order effects if the "1-n" terms are small (in a dimensionless sense) compared with pressure gradient effects. And for small values of n, the second derivative terms on the right side may be unimportant since such flows are known to contain flat velocity profiles. Crochet, Davies and Walters (1984), which deals exclusively with non-Newtonian flows, presents excellent discussions on different limit processes.

For the foregoing limits, the principal effects of nonlinearity can be modelled using the simpler and stabler Poisson model that results, one not unlike Equation 1-1 for Newtonian flow. Of course, the apparent viscosity that acts in concert with the driving pressure gradient is still variable, nonlinear, and dependent on input parameters. For such cases only - and all solutions obtained in this fashion should be checked aposteriori against the complete equation - we have

$$\partial^2 u/\partial y^2 + \partial^2 u/\partial x^2 = N(\Gamma)^{-1} \, \partial P/\partial z \qquad (2\text{-}31)$$

where Equation 2-29 is retained in its entirety. This approximation does not always apply. But the strong influence of local geometry on annular velocity (e.g., low bottom speeds in eccentric holes regardless of rheology or flow rate) suggests that any errors incurred by using Equation 2-31 may be insignificant. This simplification is akin to the "local linearization" method used in nonlinear aerodynamics. In any case, the exact geometry of the eccentric annulus is always retained.

As noted earlier in Chapter 1, borehole temperature may be important in drilling and production applications. Many studies neglect heat generation by internal friction. In temperature log interpretation using data obtained while drilling, sensor measurements may require corrections to account for fluid rheology, pumping rate and total circulation time effects. Internal heat generation may also affect local fluid viscosity since n and k depend on temperature.

One way to estimate the importance of such effects is through the strength of the temperature sources distributed within the annulus. The starting point is the energy equation for the temperature T(z,y,x,t). Even if the velocity is steady in time, temperature does not have to be. For example, in a closed system, temperatures will increase if the borehole walls do not conduct heat away as quickly as it is produced; weak heat production can lead to large increases in T over time. If these increases are significant enough, the changes of viscosity as functions of T must be modelled. This leads to mathematical complications; if the laminar viscosity $\mu = \mu(T)$ in Newtonian flow depends on temperature, say, then the momentum and energy equations will couple through this dependence.

We will not consider this coupling. We assume that all rheological input parameters are constants, so that our velocities obtain independently of T. Now the energy equation for T contains a positive definite quantity Φ called the "dissipation function", that is the distributed energy source term responsible for local heat generation. In general, it takes the form

$$\Phi(z,y,x) = S_{zz}\partial u/\partial z + S_{yy}\partial v/\partial y + S_{xx}\partial w/\partial x$$
$$+ S_{zy}(\partial u/\partial y + \partial v/\partial z) + S_{zx}(\partial u/\partial x + \partial w/\partial z)$$
$$+ S_{yx}(\partial v/\partial x + \partial w/\partial y) \tag{2-32}$$

Applying assumptions consistent with the foregoing analysis, we obtain

$$\Phi = N(\Gamma) \ \{(\partial u/\partial y)^2 + (\partial u/\partial x)^2\} > 0 \tag{2-33}$$

where, as before, we use Equation 2-29 for the apparent viscosity in its entirety. Equation 2-33 shows that velocity gradients, not magnitudes, contribute to temperature increases.

In the computed output, we will provide values of local viscous stresses and their corresponding shear rates. These stresses are the rectangular components

$$S_{zy} = N(\Gamma) \ \partial u/\partial y \tag{2-34}$$

$$S_{zx} = N(\Gamma) \ \partial u/\partial x \tag{2-35}$$

The shear rates corresponding to Equations 2-34 and 2-35 are $\partial u/\partial y$ and $\partial u/\partial x$ respectively. These quantities are useful for several reasons. They are physically important in estimating the efficiency with which fluids in deviated wells remove cuttings beds having specified mechanical properties. From the numerical analysis point of view, they allow checking of computed solutions for physical consistency (e.g., high values at solid surfaces, zeros within plug flows) and required symmetries.

We next discuss some mathematical issues related to computational grid generation and numerical solution. These ideas are highlighted because we will solve the complete boundary value problem, satisfying no-slip velocity conditions exactly, without simplifying the annular geometry.

BOUNDARY CONFORMING GRID GENERATION

In many engineering problems, a judicious choice of coordinate systems simplifies calculations and brings out the salient physical features more transparently than otherwise. For example, the use of cylindrical coordinates for single well problems in petroleum engineering leads to elegant "radial flow" results that have proven useful in well testing. Cartesian grids, on the other hand, are preferred in simulating oil and gas flows from rectangular fields.

The annular geometry modelling considered here is aimed at eccentric flows with cuttings beds, arbitrary borehole wall deformations, and unconventional drill collar or casing-centralizer cross-sections. Obviously, simple coordinate transforms are not readily available to handle arbitrary regions of flow. Without resorting to crude techniques, for instance, applying boundary conditions along mean circles and squares or invoking slot flow assumptions, there is no real reason for optimism.

Fortunately, recent developments in differential geometry allow us to construct "boundary conforming, natural coordinates" for the sake of computation. These general techniques extend classical ideas on conformal mapping. While only a decade old, they have nevertheless accelerated progress in simulating aerospace fluid mechanics, although they have yet to be applied in the petroleum industry. Thompson, Warsi and Mastin (1985) provides an excellent introduction to the subject.

To those familiar with conventional analysis, it may seem that the choice of (y,x) coordinates in Equation 2-31 is somehow unnatural. After all, in the limit of a concentric annulus, the equation does not reduce to the expected radial formulation. But our use of (y,x) coordinates was motivated by the new gridding methods which, like classical conformal mapping, are founded on Cartesian coordinates.

The approach in essence requires us to first solve a set of nonlinearly coupled, second order PDEs. In particular, the equations

$$(y_r^2 + x_r^2)\, y_{ss} - 2(y_s y_r + x_s x_r)\, y_{sr} + (y_s^2 + x_s^2)\, y_{rr} = 0 \qquad (2\text{-}36)$$

$$(y_r^2 + x_r^2)\, x_{ss} - 2(y_s y_r + x_s x_r)\, x_{sr} + (y_s^2 + x_s^2)\, x_{rr} = 0 \qquad (2\text{-}37)$$

are considered together with special mapping conditions related to the annular geometry. Equations 2-36 and 2-37 are importantly solved on simple rectangular (r,s) grids. Once the solution is obtained, the results for x(r,s) and y(r,s) are used to generate all the metric transformations needed to reformulate the physical equations for u in (r,s) coordinates. The flow problem is then solved in these rectangular computational coordinates using standard numerical methods.

We emphasize that these new coordinates implicitly contain all the details of the input geometry, providing fine resolution in tight spaces as needed. To see why, we now describe the boundary conditions used in the mapping. Figures 2-4a and 2-4b indicate how a general annular region would map into a rectangular computational space under the proposed scheme.

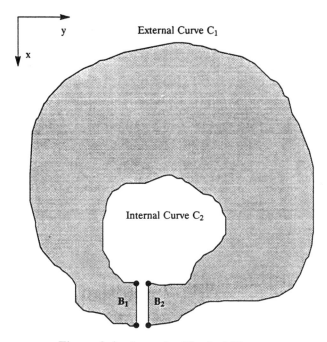

Figure 2-4a. Irregular Physical Plane.

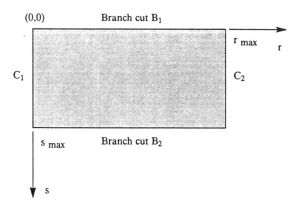

Figure 2-4b. Rectangular Computational Plane.

Again the key idea rests with the special computational coordinates (r,s). A discrete set of "user selected" physical coordinates (y,x) along curve C_1 in Figure 2-4a is specified along the straight line $r = 0$ in Figure 2-4b. Similarly, (y,x) values obtained from curve C_2 in Figure 2-4a are specified along $r = r_{max}$ in Figure 2-4b. Values for (y,x) chosen along the "branch cuts" B_1 and B_2 in Figure 2-4a are required to be single-valued along the edges $s = 0$ and $s = s_{max}$ in Figure 2-4b.

With (y,x) completely prescribed along the rectangle of Figure 2-4b, Equations 2-36 and 2-37 for y(r,s) and x(r,s) can be numerically solved. Once the solution is obtained, the one-to-one correspondences between all physical points (y,x) and computational points (r,s) are known. The latter is the domain chosen for numerical computation for the annular velocity. Finite difference representations of the no-slip conditions that apply along C_1 and C_2 of Figure 2-4a are very easily implemented in the rectangle of Figure 2-4b. At the same time, the required modifications to the governing equation for u(y,x) are only modest in scope.

Following standard coordinate transformations, the simplified Equation 2-31 becomes

$$(y_r^2 + x_r^2)\, u_{ss} - 2(y_s y_r + x_s x_r)\, u_{sr}$$
$$+ (y_s^2 + x_s^2)\, u_{rr} = (y_s x_r - y_r x_s)^2\, \partial P / \partial z\ / N(\Gamma) \tag{2-38}$$

whereas the transformed result corresponding to the Equation 2-30, not shown here, requires several additional terms. For Equation 2-38 and its more exact counterpart, the velocity terms in the apparent viscosity $N(\Gamma)$ of Equation 2-29 transform according to

$$u_y = (x_r u_s - x_s u_r)/(y_s x_r - y_r x_s) \tag{2-39}$$

$$u_x = (y_s u_r - y_r u_s)/(y_s x_r - y_r x_s) \tag{2-40}$$

These same results are used to evaluate the dissipation function Φ in Equation 2-33. It is important to emphasize that the foregoing transformations apply to all fluids, whether they are Newtonian, power law, Bingham plastic or Herschel-Bulkley. We next describe the technique used in solving Equations 2-36 through Equation 2-40.

NUMERICAL FINITE DIFFERENCE SOLUTION

We have transformed the computational problem for the annular speed u from an awkward one in the physical plane (y,x) to a simpler one in (r,s) coordinates, where the irregular domain becomes rectangular. In doing so, we introduced the intermediate problem dictated by Equations 2-36 and 2-37. When solutions for y(r,s) and x(r,s) and their corresponding metrics are available, Equation 2-38, which is slightly more complicated than the original Equation 2-31, can be solved conveniently using "rectangle based" methods without compromising the annular geometry.

Equations 2-36 and 2-37 were solved by rewriting them as a single vector equation, employing simplifications from complex variables, and discretizing the end equation using second-order accurate formulas. The finite difference equations are then reordered so that the coefficient matrix is sparse, banded and computationally efficient. Finally, the Successive Line Over Relaxation (SLOR) method was used to obtain the solution in an implicit and iterative manner. The SLOR scheme is unconditionally stable on a linearized von Neumann basis (for example, refer to Lapidus and Pinder, 1982).

Mesh generation requires approximately one minute of computing time on IBM PC-XTs equipped with math co-processors. On "386" machines, this time requirement is reduced to ten seconds. Once the transformations for y(r,s) and x(r,s) are available for a given annular geometry, Equations 2-38 to 2-40 can be solved any number of times for different applied pressure gradients, volume flow rate constraints or fluid rheology models, without recomputing the necessary mapping.

Because Equation 2-38 is similar structurally to Equations 2-36 and 2-37, the same procedure was used for its solution. These iterations converged quickly and stably because the host meshes used were physically smooth. When solutions for the velocity field u(r,s) are available (these also require one minute on XT machines), simple inverse mapping relates each computed "u" with its unique image in the physical (y,x) plane. With u(y,x) and its spatial derivatives known, post-processed quantities like $N(\Gamma)$, S_{zy}, S_{zx}, their corresponding shear rates, apparent viscosities and Φ are easily calculated and displayed in physical (y,x) coordinates.

Practitioners in drilling and production engineering recognize that flow properties within an eccentric annulus are correlatable to some extent with annular position (e.g., low bottom speeds obtained regardless of rheology). For this reason, graphical display software was developed to project u(y,x) and all post-processed quantities directly on the annular geometry. This helps visual correlation of computed physical properties or inferred characteristics (e.g., "cuttings transport efficiency" and "stuck pipe probability") with annular geometry quickly and efficiently. This highly visual output, together with the speed and stability of the numerical scheme, promotes an understanding of annular flow results in an interactive, real-time manner.

Finally, we return to the problem of fluid rheologies having non-zero yield stresses. In general there may exist internal boundaries that separate coexisting "dead", "plug" and "shear" flow regimes within the annulus. These unknown boundaries must be obtained as part of the numerical solution. In classical free surface theory for water waves, or in shock-fitting methods for gasdynamic discontinuities, explicit equations are written for the boundary curve and solved simultaneously with the field equations. These approaches, while precise, are overly complicated.

Rather, a technique similar to the popular "shock capturing" scheme used in transonic flows with embedded discontinuities was employed to develop the three zones naturally during the relaxation. The logical conditions stated in Equations 2-14b and 2-14c were added to the baseline "zero yield stress" computer model described above. This entailed a tedious "point by point" testing during the computations, where the inequalities were evaluated based upon latest available iterative solutions. But flows containing plugs generally converged faster than flows without, because fewer matrix setups and inversions (steps necessary in computing shearing motions) were required once plug and dead flow regimes developed.

EXAMPLE CALCULATIONS

We first discuss results for four special annular geometries containing nonlinear power law fluids. In particular, we consider (i) a fully concentric annular cross-section to establish a baseline reference; (ii) the same concentric combination blocked by a thick cuttings bed; (iii) a highly eccentric annular flow with the circular pipe displaced within a circular hole without a cuttings bed; and (iv) a circular borehole containing a square drill collar.

Again the contours are not restricted to circles and squares, since the algorithm efficiently solves any combination of closed curves just as easily. One set of reference flow conditions will be calculated as the basis of comparison. These comparisons allow us to validate the physical consistency of our computed results. The simplicity of the input required to run the computer program is emphasized, as well as the highly visual format of all calculated quantities. These presently include axial velocity, apparent viscosity, two rectangular components of viscous stress and shear rate, Stokes product and dissipation function.

Example 1: Fully Concentric Annular Flow

The program requests input radii and center coordinates for circles that need not be concentric. Once entered, the (y,x) coordinates of both circles are displayed in tables. If the borehole and drillpipe contours require modification, for example, to model cuttings beds or square drill collars, new coordinates entered at the keyboard replace those displayed. Alternatively, the annular contours may be "drawn" using an on-line text editor, whose results are read and interpolated by the program.

Let us consider a 2 inch radius pipe centered in a 5 inch radius borehole. Assuming the first input option, the program displays the resultant geometry and provides the user with an indication of relative dimensions, as shown in Figure 2-5a. For complete portability, all graphical input and output files appear as ASCII text characters, and do not require special computer display hardware or software.

The program checks for input errors by asking the user to verify that the pipe is wholly contained within the hole. This visual check ensures that contours do not overlap; it is important when eccentric circles are modeled or when there is significant borehole wall deformation. If the input geometry is realistic, automatic mesh generation proceeds, with all grid parameters chosen internally and transparently to the user.

The numerical mapping procedure typically requires one minute on IBM PC-XTs with math co-processors. This assumes 24 "circumferential" and 10 "radial" grids. Since the mesh system is variable, providing high resolution in tight spaces where large physical gradients are expected, and since the central difference scheme used is second-order accurate, it should more than suffice for most purposes.

When the iterations are completed, the computed mesh is displayed. The mesh constructed for our concentric annulus, shown in Figure 2-5b, is also concentric. To visualize it, simply connect points having like coordinate elevations; then sketch in orthogonals to these closed curves. Although our formulation of the mesh generation problem was undertaken in Cartesian coordinates, it is clear that the end result must be the radial grid system one naturally anticipates. The power of the method, of course, is the extension of "radial" to complicated geometries.

```
X/Y orientation:

+---> Y
|
V
  X
```

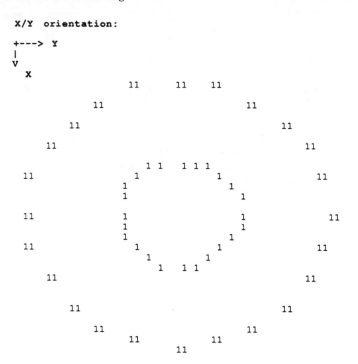

Figure 2-5a. Concentric Circular Pipe and Hole.

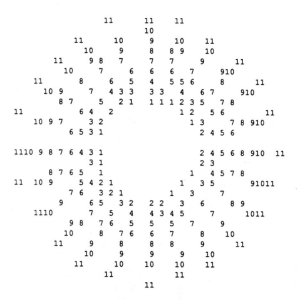

Figure 2-5b. Computed Radial Mesh System.

For a given input geometry, the mesh is determined only once; any number of physical simulations, to include changes to applied pressure gradient, total flow rate or fluid rheology, can be performed on that mesh. The program next asks if the working fluid is Newtonian, power law, Bingham plastic or Herschel-Bulkley. For Newtonian flows, only the laminar viscosity is entered; the power law mode can also be used with n = 1, but this results in slightly lengthier calculations. In the case of power law fluids, one specifies the fluid exponent n and the consistency factor k. For Bingham plastics and Herschel-Bulkley fluids, the yield stress is additionally required.

In this example, a power law fluid is assumed with an exponent of n = 0.724 and a consistency factor of .1861E-04 lbf sec^n/sq in. The axial pressure gradient is .3890E-02 psi/ft. The iterative solution to the axial velocity equation on the assumed mesh requires approximately one minute of computing time on a PC-XT. The calculated results for axial velocity, apparent viscosity, two rectangular stress and shear rate components, Stokes product and dissipation function, are always displayed on the annular geometry as shown in Figures 2-5c to 2-5g. This allows convenient correlation of physical trends with annular location.

For plotting convenience, the first two significant digits of the dependent variable in the system of units used are printed (exact magnitudes are available in tabulated output). In Figure 2-5c, the first two digits of axial velocity in "in/sec" are displayed; lines of constant velocity are obtained by connecting numbers having like values.

We emphasize again that the foregoing solution has not been "corrected" or "mesh calibrated" in the sense of Chapter 1. The computed results are displayed "as is" using internally selected mapping parameters. The objective is not so much an exact solution in the analytical sense, but accurate *comparative* solutions for a set of runs. This limited objective is more relevant to field applications.

In Figure 2-5c, the 0s found at both inner and outer circular boundaries indicate that no-slip conditions have been properly and exactly satisfied. Reference to tabulated results shows that all expected symmetries are adequately reproduced by the numerical scheme.

We emphasize that the ASCII character-based plotting routine provides only approximate results. "Missing numbers" and lack of symmetry are due to decimal truncation, array normalization, and character and line spacing issues. The plotter is intended as an inexpensive visualization tool that is universally portable. For more precise displays, commercial software and hardware packages are recommended.

The exact results in any event are available in output files. Typical results for the axial speed U are shown in Table 2-1. The "circumferential grid block index" is given in the left column, and corresponding coordinates appear just to the right. The index takes on a value of "1" at the "bottom middle" of any particular closed contour, and increases clockwise to "24".

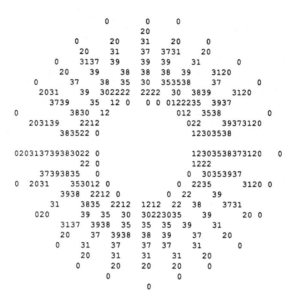

Figure 2-5c. Annular Velocity.

The printed results in Table 2-1 for Contours No. 1 and 11 demonstrate that "no-slip" conditions are rigorously enforced at all solid boundaries. Here, the maximum speeds are found along Contour No. 6 (note the 38s and 39s in Figure 2-5c). Since the flow is concentric, all U's along this contour must be identical; this is always satisfied to the third decimal place. For internal nodes 4 to 20, the Us are identical to four places. The computed "as is" annular volume flow rate is 457.8 gal/min.

Computed results for the apparent viscosity are plotted in Figure 2-5d; representative values are tabulated in Table 2-2. Unlike Newtonian flows, the apparent viscosity in power law flows varies with position, and changes from problem to problem. For the present example, it is largest near the center of the annulus.

Table 2-1
Example 1:   Annular Velocity (in/sec)

Results for pipe/collar boundary, Contour No. 1:
1 X= .8000E+01 Y= .6000E+01 U= .0000E+00
2 X= .7932E+01 Y= .5482E+01 U= .0000E+00
3 X= .7732E+01 Y= .5000E+01 U= .0000E+00
4 X= .7414E+01 Y= .4586E+01 U= .0000E+00

Results for Contour No. 6:
1 X= .9149E+01 Y= .6000E+01 U=-.3873E+02
2 X= .9045E+01 Y= .5184E+01 U=-.3873E+02
3 X= .8732E+01 Y= .4423E+01 U=-.3874E+02
4 X= .8231E+01 Y= .3769E+01 U=-.3875E+02
5 X= .7578E+01 Y= .3267E+01 U=-.3875E+02
6 X= .6817E+01 Y= .2952E+01 U=-.3875E+02

Result for borehole annular boundary, Contour No. 11:
1 X= .1100E+02 Y= .6000E+01 U= .0000E+00
2 X= .1083E+02 Y= .4706E+01 U= .0000E+00
3 X= .1033E+02 Y= .3500E+01 U= .0000E+00
4 X= .9536E+01 Y= .2464E+01 U= .0000E+00

Results for the viscous stresses "Apparent Viscosity x dU(y,x)/dx" and "Apparent Viscosity x dU(y,x)/dx" are presented in Figures 2-5e and 2-5f and Tables 2-3 and 2-4. The plotting routine omits the signs of these stresses for visual clarity; signs and exact magnitudes are available from tabulated results.

This test case assuming concentric flow is important for numerical validation. The dU(y,x)/dx stress is symmetric with respect to the horizontal center line, as required, and vanishes there; similarly, the dU(y,x)/dy stress is symmetric with respect to the vertical center line and is zero there. These physically correct results appear as the result of properly converged iterations. Finally, results for the dissipation function are shown in Figure 2-5g and Table 2-5. Note how the greatest heat generation occurs at the pipe surface and at the borehole wall; there is minimal dissipation at the midpoint of the annulus.

```
                5         5    5
                          6
          5          6        7      6       5
               6        7        9    9 7      6
        5       7 9   15      15   15      7           5
            6     15     11    11   11   15      7 6
      5         9     11    8     7    8 811      9         5
        6 7    15     7 6 6    6 6    7  1115      7 6
        915      8     6 5    5 5 5 6 6 8   15 9
    5         11 7    6               5 6     811              5
      6 715      6 6                  5 6      15 9 7 6
         11 8 6 5                      6 7 811

   5 6 7 91511 7 6 5                      6 7 811 9 7 6    5
            6 5                          6 6
         91511 8    5                  5    7 815 9
   5    6 7      8 7 6 5              5    6 8        7 6 5
        1511     6 6 5            5    6     15
        7      11 8    6 6       6 6    6 11      9 7
     5 6        15     8    7    7 6 7 8     15          6 5
            7 9  1511    8      8    8 15      7
            6       9  1511   11   15      9        6
            5      7       9    9  9      7        5
               6       7    7      7      6
            5          6    6      6      5
               5              5
                          5
```

Figure 2-5d. Apparent Viscosity.

```
                29        30     29
                          25
               26      24     17     24      26
                  22     17     10    917      22
         21     15 8    2      2    2     15        21
            18      2      6    6    6    2      1218
        15       7      5  13    20  1311 5      7          15
         12 8      2   172326   2726   17    4 2       812
           5 1       9   2938   3938342919 9    1 5
     7          310  24             2824    6 3              7
       6 4 0      1317               1913        0 2 4 6
          1 3  710                    8 5 3 1

   0 0 0 0 0 0 0 0 0                       0 0 0 0 0 0    0
               710                         8 7
        2 0 1 3  19                     19    5 3 0 2
     7    6 4      6101728               28   13 6        4 6 7
          1 3  192434               34   19      1
         8       4 9  2333   3633   23    4      5 8
      1512       2   11   19   22261911     2          1215
         12 7     2 5   13   14   13    2      12
         18       8    2 6    7    2      8     18
        21      15      9    10    9    15      21
             22     17     18    17     22
           26      24     27     24      26
                  29               29
                          32
```

Figure 2-5e. Stress "AppVisc × dU(y,x)/dx".

```
                    7         0     7
                              0
          15          6       0       6     15
              12      4       0   2 4     12
      21        8 5   0       0   0       8       21
          18      1     1     0   1   1       1218
      26        7       3   3 0   3 6 3     7       26
        2215      2   1013 7   0 7  10    4 2     1522
          8 2       9  1710   010191719 9   2 8
    29          517  24               2824  11 5               29
        2417 2    2329               3423        2 91724
          6132638                   331913 6

  30251710 2 6202739                342013 6101725  30
              2638                   3326
          9 2 613  34               34  1913 2 9
    29   2417    11172928            28  2311        172429
          2 5  192419               19  19      2
        15      4 9  13 8   0 8  13    4       815
      2622        2   6   5   0 7 5 6     2       2226
          12 7   1 3   3   0   3   1     12
          18      5   0 1   0   0     5   18
        21      8       2   0   2     8       21
          12        4     0     4   12
          15        6     0     6   15
              7               7
                              0
```

Figure 2-5f. Stress "AppVisc × dU(y,x)/dy".

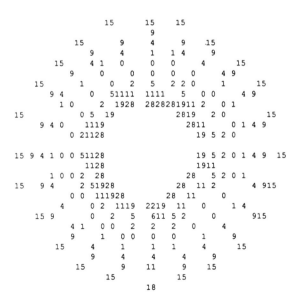

Figure 2-5g. Dissipation Function.

Table 2-2
Example 1:   Apparent Viscosity (lbf sec/sq in)

Results for pipe/collar boundary, Contour No. 1:
5 X= .7000E+01 Y= .4268E+01 AppVisc= .5558E-05
6 X= .6518E+01 Y= .4068E+01 AppVisc= .5558E-05
7 X= .6000E+01 Y= .4000E+01 AppVisc= .5558E-05
8 X= .5482E+01 Y= .4068E+01 AppVisc= .5558E-05
9 X= .5000E+01 Y= .4268E+01 AppVisc= .5558E-05
10 X= .4586E+01 Y= .4586E+01 AppVisc= .5558E-05
11 X= .4268E+01 Y= .5000E+01 AppVisc= .5558E-05
           Results for Contour No.   7:
7 X= .6000E+01 Y= .2541E+01 AppVisc= .1577E-04
8 X= .5105E+01 Y= .2659E+01 AppVisc= .1577E-04
9 X= .4271E+01 Y= .3005E+01 AppVisc= .1577E-04
Result for borehole annular boundary, Contour No. 11:
9 X= .3500E+01 Y= .1670E+01 AppVisc= .5790E-05
10 X= .2464E+01 Y= .2464E+01 AppVisc= .5790E-05
11 X= .1670E+01 Y= .3500E+01 AppVisc= .5790E-05
12 X= .1170E+01 Y= .4706E+01 AppVisc= .5790E-05
13 X= .1000E+01 Y= .6000E+01 AppVisc= .5790E-05
14 X= .1170E+01 Y= .7294E+01 AppVisc= .5790E-05

Table 2-3
Example 1:   Stress "AppVisc x dU(y,x)/dx" (psi)

Results for pipe/collar boundary, Contour No. 1:
1 X= .8000E+01 Y= .6000E+01 Stress=-.4267E-03
2 X= .7932E+01 Y= .5482E+01 Stress=-.3864E-03
3 X= .7732E+01 Y= .5000E+01 Stress=-.3460E-03
           Results for Contour No.   6:
4 X= .8231E+01 Y= .3769E+01 Stress=-.4616E-04
5 X= .7578E+01 Y= .3267E+01 Stress=-.3260E-04
6 X= .6817E+01 Y= .2952E+01 Stress=-.1686E-04
Result for borehole annular boundary, Contour No. 11:
11 X= .1670E+01 Y= .3500E+01 Stress=-.2660E-03
12 X= .1170E+01 Y= .4706E+01 Stress=-.2967E-03
13 X= .1000E+01 Y= .6000E+01 Stress=-.3072E-03
14 X= .1170E+01 Y= .7294E+01 Stress=-.2967E-03

Table 2-4
Example 1:   Stress "AppVisc x dU(y,x)/dy" (psi)

Results for pipe/collar boundary, Contour No. 1:
8 X= .5482E+01 Y= .4068E+01 Stress= .3858E-03
9 X= .5000E+01 Y= .4268E+01 Stress= .3459E-03
10 X= .4586E+01 Y= .4586E+01 Stress= .2824E-03
Results for Contour No. 6:
20 X= .6817E+01 Y= .9048E+01 Stress=-.6305E-04
21 X= .7578E+01 Y= .8733E+01 Stress=-.5656E-04
22 X= .8231E+01 Y= .8231E+01 Stress=-.4622E-04
23 X= .8732E+01 Y= .7577E+01 Stress=-.3270E-04
24 X= .9045E+01 Y= .6816E+01 Stress=-.1681E-04
Result for borehole annular boundary, Contour No. 11:
1 X= .1100E+02 Y= .6000E+01 Stress=-.5091E-05
2 X= .1083E+02 Y= .4706E+01 Stress=-.7929E-04
3 X= .1033E+02 Y= .3500E+01 Stress=-.1535E-03

Table 2-5
Example 1:   Dissipation Function (lbf/(sec x sq in))

Results for pipe/collar boundary, Contour No. 1:
6 X= .6518E+01 Y= .4068E+01 DissipFn= .2870E-01
7 X= .6000E+01 Y= .4000E+01 DissipFn= .2870E-01
8 X= .5482E+01 Y= .4068E+01 DissipFn= .2870E-01
Results for Contour No. 7:
9 X= .4271E+01 Y= .3005E+01 DissipFn= .5235E-04
10 X= .3554E+01 Y= .3554E+01 DissipFn= .5236E-04
11 X= .3005E+01 Y= .4271E+01 DissipFn= .5236E-04
12 X= .2659E+01 Y= .5105E+01 DissipFn= .5237E-04
Result for borehole annular boundary, Contour No. 11:
12 X= .1170E+01 Y= .4706E+01 DissipFn= .1629E-01
13 X= .1000E+01 Y= .6000E+01 DissipFn= .1629E-01
14 X= .1170E+01 Y= .7294E+01 DissipFn= .1629E-01
15 X= .1670E+01 Y= .8500E+01 DissipFn= .1629E-01

Example 2:   Concentric Pipe and Borehole
in the Presence of a Cuttings Bed

For comparative purposes, we consider the same annular geometry used previously; that is, a 2 inch radius pipe located within a concentric 5 inch radius borehole. But when the program requests modifications to the outer contour, we overwrite five of the bottom coordinates to simulate a flat cuttings bed. The bed height is approximately one-half of the distance up the annular cross-section. This area blockage should reduce the "457.8 gal/min" obtained in Example 1 for the unblocked annulus.

First, the program generates the grid shown in Figure 2-6a, which conforms to the top of the cuttings bed. To compare our results with Example 1, we again assume a power law fluid with an exponent n = 0.7240 and a consistency factor of .1861E-04 lbf sec^n/sq in. The pressure gradient is the same .3890E-02 psi/ft. Figure 2-6b shows that the maximum velocities at the bottom are less than one-half of those at the top. This trend is well known qualitatively, but the program allows us to obtain exact velocities everywhere without making unrealistic "slot flow" assumptions. The low velocities adjacent to the bed imply that cuttings transport will not be very efficient just above it, and that stuck pipe is possible.

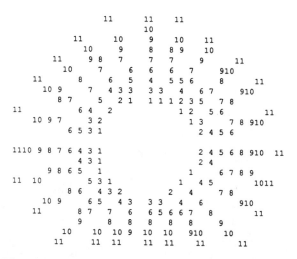

Figure 2-6a. Mesh System for Hole with Cuttings Bed.

```
                 0        0    0
                         20
         0       20      31   20     0
            20      31   37  3731   20
      0    3137   39     39   39   31          0
         20    39    38   38  38   39    3120
      0      37    38  35   30  353538   37        0
     2031    39  302222  2222  30  3839     3120
     3738    35  12 0    0 0 0122235  3837
   0       3830  12            012   3538             0
   203138    2212              022     38373120
      383522 0                12293538

020313638372921 0             11293437363120   0
        2820 0               1128
   31363633   0             0    36373631
 0   20      3119 0           0  2631      20 0
     3534  2215  6         6  22     3535
    2030    2926 15 8    8 8  15  29       3020
  0      3031  21  15  1514152121  30           0
       27    20  14   14  14  20    27
     19    10   611   6   6  1710    19
   0     0    0     0   0    0      0
```

Figure 2-6b. Annular Velocity.

```
                   5        5    5
                            6
          5         6       7    6      5
             6        7       9   9 7     6
         5      7 9  15     15  15     7         5
           6    15    11   11  11  15      7 6
        5      9    11   8    7   8 811      9        5
       6 7    15    7 6 6    6 6    7  1115     7 6
       915      8    6 5   5 5 5 6 6 8  15 9
   5        11 7    6          5 6    811             5
     6 715    6 6                5 6    15 9 7 6
       11 8 6 5                  6 7 811

5 6 7 91511 7 6 5                  6 7 811 9 7 6   5
        7 6 5                     6 7
     81011 8    5                 5   111510 8
5   6         8 6 5               5   7 8        6 5
      1010    7 6 6             6   7   1310
     7 8      9 8   7 7     7 7   7   9        8 7
    5       910    9   11   14 911 9 9    9          5
         8        8   9    9   9   8     8
        7      6   7 8    7   7   7 6      7
        6      6   6     6   6    6        6
```

Figure 2-6c. Apparent Viscosity.

```
                        29      30     29
                                25
              26        24     17    24     26
                 22     17     10   917    22
        21     15 8  2         2    2    15        21
           18      2    6      6    6    2      1218
      15          7     5  13  20  1311 5      7          15
         12 8        1  172326 2726 17   4 1        812
           4 1       9  2938 3938342919 9   1 4
    7              310 24                2824    6 3              7
       6 4 0      1317                   1913       0 2 4 6
          2 3 710                         8 5 3 2

  0 0 0 0 1 0 0 0 0                       0 0 0 0 0 0 0   0
            4 6 9                          8 4
       5 3 0 2  18                        18       0 2 3 5
    7    6          41124                 24    8 4          6 7
            7 0  111522                   22   11     3 7
       1310        0 6  1219  2019 12   0         1013
    15          10 4    6   2   5 8 2 0 6  10            15
              14      12  10   8   10  12      14
            19      24  2115  17  21  1824   19
          22        28  25    20  25      28        22
```

Figure 2-6d. Stress "AppVisc × dU(y,x)/dx".

```
                      7         0      7
                                0
              15           6    0     6    15
                 12        4    0   2 4    12
          21      8 5   0        0   0      8         21
             18       1     1    0   1   1      1218
       26       7       3   3    0   3 6 3      7        26
        2215      2  1013 7      0 7  10    4 2    1522
          8 2       9  1710      010191719 9    2 8
    29            517  24                2824  11 5            29
      2415 2    2329                      3423      2 91524
         6132638                          331913 6

  302517 9 2 6202739                       342013 6 91725  30
          202637                           3220
       15 8 713  32                        32      7 0 815
   28   23        132325                   25  1813          2328
          4 8  151914                      14  15     2 4
       1911       1013  12 5    3 5  12  10         1119
       23          2 7   7   5   2 5 510 7    2              23
              1       5   3    0   3 5       1
                6     1  1 2    1   1  2 1     6
              9       0   0    1   0      0       9
```

Figure 2-6e. Stress "AppVisc × dU(y,x)dy".

```
                    15        15    15
                                9
              15         9     4      9    15
                    9      4   1    1 4     9
        15       4 1    0     0    0      4      15
           9       0      0   0    0    0      4 9
    15         1      0    2   5    2 2 0     1         15
        9 3      0   51111 1111   5    0 0    3 9
        1 0      2  1928 2828281911 2    0 1
 15            0 5  19              2819  2 0              15
     9 3 0      1119              2811     0 0 3 9
          0 21128                 19 5 2 0

 15 9 3 0 0 0 51127              18 5 2 0 0 3 9  15
             51126              18 5
        3 0 0 2  24               24      0 0 0 3
 15    8          21021           21    5 2           815
            0 0    5 911         11    5    0 0
         7 2      1 2    4 5   5 5   4    1       2 7
     13           1 0    1    0   0 1 0 1 1   1         13
                  2      2   1   0   1   2      2
            5         8    6 3   4   6    4 8     5
            9         13   9     6   9    13      9
```

Figure 2-6f. Dissipation Function.

The program can also be used to determine the mud type needed to increase bottom velocities or viscous stresses to acceptable levels. The computed "as is" volume flow rate is 366.7 gal/min, much less than the 457.8 gal/min obtained for the unblocked flow of Example 1. This decreased value is consistent with physical intuition. Numerical results along the vertical line of symmetry are given in Table 2-6. These numbers represent upper and lower velocity profiles; again, note the exact implementation of no-slip conditions.

Table 2-6
Example 2:   Annular Velocity (in/sec)

Velocity Profile (Vertical "X" coordinate increases downward)
Upper Annulus
\# 13 X= .1000E+01 Y= .6000E+01 U= .0000E+00
\# 13 X= .1441E+01 Y= .6000E+01 U=-.2016E+02
\# 13 X= .1843E+01 Y= .6000E+01 U=-.3189E+02
\# 13 X= .2209E+01 Y= .6000E+01 U=-.3756E+02
\# 13 X= .2541E+01 Y= .6000E+01 U=-.3917E+02
\# 13 X= .2844E+01 Y= .6000E+01 U=-.3871E+02
\# 13 X= .3120E+01 Y= .6000E+01 U=-.3590E+02
\# 13 X= .3372E+01 Y= .6000E+01 U=-.3050E+02
\# 13 X= .3601E+01 Y= .6000E+01 U=-.2259E+02
\# 13 X= .3810E+01 Y= .6000E+01 U=-.1235E+02
\# 13 X= .4000E+01 Y= .6000E+01 U= .0000E+00
Lower Annulus
\# 1 X= .8000E+01 Y= .6000E+01 U= .0000E+00
\# 1 X= .8096E+01 Y= .6000E+01 U=-.4868E+01
\# 1 X= .8202E+01 Y= .6000E+01 U=-.8927E+01
\# 1 X= .8319E+01 Y= .6000E+01 U=-.1207E+02
\# 1 X= .8447E+01 Y= .6000E+01 U=-.1420E+02
\# 1 X= .8588E+01 Y= .6000E+01 U=-.1528E+02
\# 1 X= .8744E+01 Y= .6000E+01 U=-.1532E+02
\# 1 X= .8914E+01 Y= .6000E+01 U=-.1414E+02
\# 1 X= .9102E+01 Y= .6000E+01 U=-.1140E+02
\# 1 X= .9309E+01 Y= .6000E+01 U=-.6785E+01
\# 1 X= .9536E+01 Y= .6000E+01 U= .0000E+00

Fluid viscosity is important to cuttings transport, and plays a dominant role in near-vertical holes. For example, in Newtonian flows the drag force acting on a slowly moving small particle is proportional to the product of viscosity and the relative speed. In power law flows, the apparent viscosity varies with space, but a similar correlation may apply. The apparent viscosity (among other parameters) is a useful qualitative indicator of cuttings mobility. Figure 2-6c shows the distribution of apparent viscosity computed for this problem, and Table 2-7 gives typical numerical values along the outer borehole/cuttings bed contour.

Table 2-7
Example 2:   Apparent Viscosity (lbf sec/sq in)

Result for borehole annular boundary, Contour No. 11:
```
#  1  X= .9536E+01  Y= .6000E+01  AppVisc= .6895E-05
#  2  X= .9536E+01  Y= .5000E+01  AppVisc= .6545E-05
#  3  X= .9536E+01  Y= .4000E+01  AppVisc= .6196E-05
#  4  X= .9535E+01  Y= .2464E+01  AppVisc= .6421E-05
#  5  X= .8500E+01  Y= .1670E+01  AppVisc= .5875E-05
#  6  X= .7294E+01  Y= .1170E+01  AppVisc= .5793E-05
#  7  X= .6000E+01  Y= .1000E+01  AppVisc= .5789E-05
#  8  X= .4706E+01  Y= .1170E+01  AppVisc= .5791E-05
#  9  X= .3500E+01  Y= .1670E+01  AppVisc= .5791E-05
# 10  X= .2464E+01  Y= .2464E+01  AppVisc= .5791E-05
# 11  X= .1670E+01  Y= .3500E+01  AppVisc= .5791E-05
# 12  X= .1170E+01  Y= .4706E+01  AppVisc= .5791E-05
# 13  X= .1000E+01  Y= .6000E+01  AppVisc= .5791E-05
# 14  X= .1170E+01  Y= .7294E+01  AppVisc= .5791E-05
# 15  X= .1670E+01  Y= .8500E+01  AppVisc= .5791E-05
# 16  X= .2464E+01  Y= .9535E+01  AppVisc= .5791E-05
# 17  X= .3500E+01  Y= .1033E+02  AppVisc= .5791E-05
# 18  X= .4706E+01  Y= .1083E+02  AppVisc= .5791E-05
# 19  X= .6000E+01  Y= .1100E+02  AppVisc= .5789E-05
# 20  X= .7294E+01  Y= .1083E+02  AppVisc= .5793E-05
# 21  X= .8500E+01  Y= .1033E+02  AppVisc= .5875E-05
# 22  X= .9535E+01  Y= .9535E+01  AppVisc= .6421E-05
# 23  X= .9536E+01  Y= .8000E+01  AppVisc= .6196E-05
# 24  X= .9536E+01  Y= .7000E+01  AppVisc= .6545E-05
```

Axial velocity and apparent viscosity play direct roles in the dynamics of individual cuttings in near-vertical holes. The stability of the beds formed by particles that have descended to the lower side of the annulus in horizontal or highly deviated holes is also of interest.

In this respect, the fluid shear stresses acting at the surface of the cuttings bed are important. If they exceed the bed yield stress, then it is likely that the bed will erode. The program can be used to determine which powers and consistency factors are needed to erode beds with known mechanical properties. There are

Table 2-8
Example 2:   Stress "AppVisc x dU(y,x)/dx" (psi)

Result for borehole annular boundary, Contour No. 11:
```
#  1  X= .9536E+01  Y= .6000E+01  Stress= .2068E-03
#  2  X= .9536E+01  Y= .5000E+01  Stress= .2508E-03
#  3  X= .9536E+01  Y= .4000E+01  Stress= .2888E-03
#  4  X= .9535E+01  Y= .2464E+01  Stress= .2289E-03
#  5  X= .8500E+01  Y= .1670E+01  Stress= .1501E-03
#  6  X= .7294E+01  Y= .1170E+01  Stress= .7748E-04
#  7  X= .6000E+01  Y= .1000E+01  Stress= .1265E-06
#  8  X= .4706E+01  Y= .1170E+01  Stress=-.7871E-04
#  9  X= .3500E+01  Y= .1670E+01  Stress=-.1528E-03
# 10  X= .2464E+01  Y= .2464E+01  Stress=-.2166E-03
# 11  X= .1670E+01  Y= .3500E+01  Stress=-.2656E-03
# 12  X= .1170E+01  Y= .4706E+01  Stress=-.2964E-03
# 13  X= .1000E+01  Y= .6000E+01  Stress=-.3069E-03
# 14  X= .1170E+01  Y= .7294E+01  Stress=-.2964E-03
# 15  X= .1670E+01  Y= .8500E+01  Stress=-.2656E-03
# 16  X= .2464E+01  Y= .9535E+01  Stress=-.2166E-03
# 17  X= .3500E+01  Y= .1033E+02  Stress=-.1528E-03
# 18  X= .4706E+01  Y= .1083E+02  Stress=-.7871E-04
# 19  X= .6000E+01  Y= .1100E+02  Stress= .1265E-06
# 20  X= .7294E+01  Y= .1083E+02  Stress= .7748E-04
# 21  X= .8500E+01  Y= .1033E+02  Stress= .1501E-03
# 22  X= .9535E+01  Y= .9535E+01  Stress= .2289E-03
# 23  X= .9536E+01  Y= .8000E+01  Stress= .2888E-03
# 24  X= .9536E+01  Y= .7000E+01  Stress= .2507E-03
```

----

two relevant components of viscous fluid stress, namely, "Apparent Viscosity x dU(y,x)/dx" and "Apparent Viscosity x dU(y,x)/dy".

Computed results for the absolute value of stress are plotted in Figures 2-6d and 2-6e; actual stresses along the borehole/cuttings bed contour are explicitly given in Tables 2-8 and 2-9. In Figure 2-6d, the weak symmetry about the horizontal row of zeros indicates that the influence of the bed is a local one (the "zeros" actually contain unprinted fractional values). But although bed effects are local in this sense, they do affect total flow rate significantly, as is known experimentally and computed here.

Table 2-9
Example 2: Stress "AppVisc x dU(y,x)/dy" (psi)

Result for borehole annular boundary, Contour No. 11:

```
#  1  X= .9536E+01  Y= .6000E+01  Stress= .1115E-04
#  2  X= .9536E+01  Y= .5000E+01  Stress= .3537E-05
#  3  X= .9536E+01  Y= .4000E+01  Stress=-.3319E-05
#  4  X= .9535E+01  Y= .2464E+01  Stress=-.9696E-04
#  5  X= .8500E+01  Y= .1670E+01  Stress=-.2353E-03
#  6  X= .7294E+01  Y= .1170E+01  Stress=-.2845E-03
#  7  X= .6000E+01  Y= .1000E+01  Stress=-.3012E-03
#  8  X= .4706E+01  Y= .1170E+01  Stress=-.2937E-03
#  9  X= .3500E+01  Y= .1670E+01  Stress=-.2646E-03
# 10  X= .2464E+01  Y= .2464E+01  Stress=-.2166E-03
# 11  X= .1670E+01  Y= .3500E+01  Stress=-.1533E-03
# 12  X= .1170E+01  Y= .4706E+01  Stress=-.7943E-04
# 13  X= .1000E+01  Y= .6000E+01  Stress=-.4881E-10
# 14  X= .1170E+01  Y= .7294E+01  Stress= .7943E-04
# 15  X= .1670E+01  Y= .8500E+01  Stress= .1533E-03
# 16  X= .2464E+01  Y= .9535E+01  Stress= .2166E-03
# 17  X= .3500E+01  Y= .1033E+02  Stress= .2646E-03
# 18  X= .4706E+01  Y= .1083E+02  Stress= .2937E-03
# 19  X= .6000E+01  Y= .1100E+02  Stress= .3012E-03
# 20  X= .7294E+01  Y= .1083E+02  Stress= .2845E-03
# 21  X= .8500E+01  Y= .1033E+02  Stress= .2353E-03
# 22  X= .9535E+01  Y= .9535E+01  Stress= .9696E-04
# 23  X= .9536E+01  Y= .8000E+01  Stress= .3319E-05
# 24  X= .9536E+01  Y= .7000E+01  Stress=-.3537E-05
```

Finally, Figure 2-6f displays the dissipation function as it varies in the annular cross-section. At least in this example, the lower part of the annulus near the cuttings bed is relatively nondissipative. Typical calculated results are given in Table 2-10.

Table 2-10
Example 2:   Dissipation Function (lbf/(sec x sq in))

Results for pipe/collar boundary, Contour No. 1:
```
#  1  X= .8000E+01  Y= .6000E+01  DissipFn= .1157E-01
#  2  X= .7932E+01  Y= .5482E+01  DissipFn= .1249E-01
#  3  X= .7732E+01  Y= .5000E+01  DissipFn= .1597E-01
#  4  X= .7414E+01  Y= .4586E+01  DissipFn= .2124E-01
#  5  X= .7000E+01  Y= .4268E+01  DissipFn= .2488E-01
#  6  X= .6518E+01  Y= .4068E+01  DissipFn= .2675E-01
#  7  X= .6000E+01  Y= .4000E+01  DissipFn= .2766E-01
#  8  X= .5482E+01  Y= .4068E+01  DissipFn= .2813E-01
#  9  X= .5000E+01  Y= .4268E+01  DissipFn= .2839E-01
# 10  X= .4586E+01  Y= .4586E+01  DissipFn= .2853E-01
# 11  X= .4268E+01  Y= .5000E+01  DissipFn= .2860E-01
# 12  X= .4068E+01  Y= .5482E+01  DissipFn= .2864E-01
# 13  X= .4000E+01  Y= .6000E+01  DissipFn= .2865E-01
# 14  X= .4068E+01  Y= .6518E+01  DissipFn= .2864E-01
# 15  X= .4268E+01  Y= .7000E+01  DissipFn= .2860E-01
# 16  X= .4586E+01  Y= .7414E+01  DissipFn= .2853E-01
# 17  X= .5000E+01  Y= .7732E+01  DissipFn= .2839E-01
# 18  X= .5482E+01  Y= .7932E+01  DissipFn= .2813E-01
# 19  X= .6000E+01  Y= .8000E+01  DissipFn= .2766E-01
# 20  X= .6518E+01  Y= .7932E+01  DissipFn= .2675E-01
# 21  X= .7000E+01  Y= .7732E+01  DissipFn= .2488E-01
# 22  X= .7414E+01  Y= .7414E+01  DissipFn= .2124E-01
# 23  X= .7732E+01  Y= .7000E+01  DissipFn= .1597E-01
# 24  X= .7932E+01  Y= .6518E+01  DissipFn= .1249E-01
```

Example 3: Highly Eccentric Circular Pipe and Borehole

We now refer to the concentric geometry of Example 1, but displace the pipe downward by 2 inches (the pipe and borehole radii are 2 and 5 inches, respectively). The program allows us to add cuttings beds and general wall deformations by modifying the boundary coordinates as before. For comparative purposes, we will not do so, thus leaving the cross-sectional areas here and in Example 1 identical. The grid selected by the computer analysis is shown below in Figure 2-7a.

We again assume a power law fluid with an exponent of n = 0.7240 and a consistency factor of .1861E-04 lbf sec^n/sq in. The axial pressure gradient is still .3890E-02 psi/ft. For brevity, we will omit tabulated results, since they are similar to those computed in Examples 1 and 2.

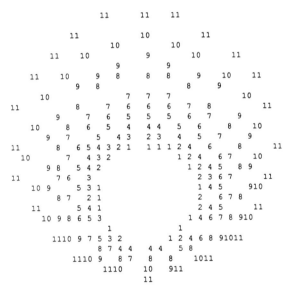

Figure 2-7a. Mesh System for Eccentric Circles.

The annular volume flow rate for this geometry is 883.7 gal/min, significantly higher than the 457.8 gal/min obtained for the concentric flow of Example 1. Again, this large increase is consistent with experimental observations indicating that higher eccentricity increases flow rates. The plots shown in Figures 2-7b to 2-7f for axial velocity, apparent viscosity, stress and dissipation function should be compared with earlier figures; the comparison reveals the qualitative and quantitative differences between the three annular geometries considered so far.

As an additional check, we evaluated the foregoing geometry with all input parameters unchanged, except that the fluid exponent is now increased to 1.5. The fluid, becoming dilatant instead of pseudoplastic, should possess a narrower velocity profile and consequently support less volume flow. Figure 2-7g displays the axial velocity solution computed. The calculated annular volume flow rate of 149.5 gal/min is much smaller, as required, than the 883.7 gal/min computed above.

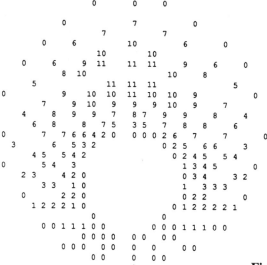

Figure 2-7b. Annular Velocity.

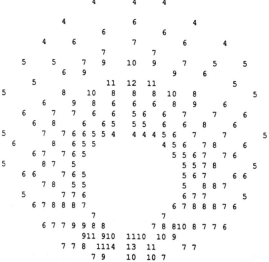

Figure 2-7c. Apparent Viscosity.

```
            41      42    41

      37              35          37
          34              34
    30    30          21        30      30
          20          20
 20   23   17    8     9    8   17    23    20
    11    5                  5    11
   14                 6    5    6                 14
9          1     7   15  15   15   7    1             9
     4        9   17  26  26   26  17    9     4
   4     4   17  25  35   3635  25  17     4     4
   2  12      24  3345    5745  33  24  12      2
1      9   182129344862  6562544829   18     9        1
   5      14   242639              423924  1714     5
    913   181827                 28271818  13 9
11       1515   17                 15171515             11
   1214     14 813                 131114       1412
     1515    4 0                    4  121515
 18          8 4 8                     4 4 8       18
   171515 8 4 5                   13 0 812151517
            15                    15
    201815 9 0 812            1412 4 412151820
          9 6 6 7        8 7    2 9
      181512    6 3      1    6     1518
            1412             6    914
                            8
```

Figure 2-7d. Stress "AppVisc × dU(y,x)/dx".

```
            11       0     11

       22            0           22
            10             10
    31     20         0       20        31
            7         7
 38    28    15    5  0    5    15    28    38
       21   11             11    21
    33               3  0   3               33
41        14     5   0  0    0    5   14         41
    24       7   0   3  0    3    0    7    24
 35    15    1   6   6  0 6   6    1    15    35
    25    7     9  12 9 0 9  12    9    7    25
41      15   31217252515 015302517   3     15     41
  34        5   213034          413421    4 5    34
   2314   142339              46392314  1423
36       3 7  32                 3932 7 3        36
  2919     153046                462315    1929
    10 0  3641                    36    8 010
 28         152133                302115        28
   2213 5 81320                  2417 8 1 51322
            15                   15
    1813 7 31013 8            8 812 7 1 71318
          0 3 7 3       2 3    7 0
      9 6 3   0 1       0    0     6 9
            3 2         3    0 3
                        4
```

Figure 2-7e. Stress "AppVisc × dU(y,x)/dy".

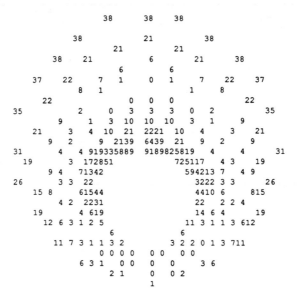

Figure 2-7f. Dissipation Function.

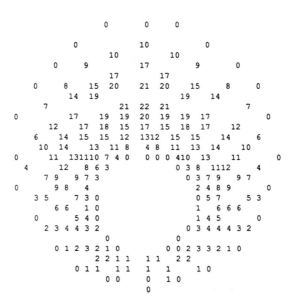

Figure 2-7g. Annular Velocity.

Example 4: Square Drill Collar in a Circular Hole

```
                    0       0   0
                    15      15  15
              0     25      26  25     0
              14    30      31  30   2214
        0     27 32      33  32    27   12 0
       20     29 3230    31  32  2929 2420
   0     26    27  26    27  26  27 2526        0
   14       23 17  11    12  11 9171423    2214
     2729     7 0    0    0 0    0   7 292927
0        2717 0                      92327            0
 1525303232                                    32302515
       302611 0                        112030

015263133312712 0                      1221313333312615 0
            0                                  0
     3032322611                          1120303230
01525        9 0                         9        2515 0
     292717                               2329
   222729  14 7 9   10   10 010 9  14      2922
   014    2523 22  18    24  18  22  25       14 0
     2425 2826 28    28  28  28   24
    1220     28 30    30  30   28  12
      0    2226 29    29  29   22          0
           14    24    24   24  14
            0    15    15   15   0
                  0        0
                    0
```

Figure 2-8a. Annular Velocity.

Square drill collars are sometimes used to control drillpipe sticking and dogleg severity. In this final comparative example, we consider a 5 inch radius borehole containing a centered square drill collar having 4 inch sides. These relative dimensions are selected for plotting purposes only. In all cases, the finite difference program will accurately calculate and tabulate output quantities, even if the ASCII plotter lacks sufficient spatial resolution.

We will assume the same flow parameters as Example 1, where we had obtained peak annular velocities of 39 in/sec and a total flow rate of 457.8 gal/min. But here, tabulated results show peak velocities of 33 in/sec obtained near the center of the annulus; also, the total volume flow rate is 334.7 gal/min. The corresponding velocity plot is shown in Figure 2-8a above. These rates are smaller than those of Example 1 because the "4 inch square" blocks more area than its inscribed "4 inch diameter circle". Thus computed results are consistent with behavior expected on physical grounds. Figures 2-8b to 2-8e are given without discussion; note how the outline of the square drill collar is adequately represented throughout.

```
                    6       5       6
                      6     6     6
              6           7     7     7       6
                7       9       9   9       7 7
          6         9  12      13  12       9       7 6
              8      12  12 9     9  12  1112     9 8
        6        15       9   8   8   8   9  1115          6
          7          8   7   6   6   6 6 7 6 8       7 7
          912        5 5   6   5 6   5     5  1112 9
      6           9 7 5                   6 7 9               6
        6 7 91212                              12 9 7 6
            9 8 6 6                         6 7 9

    5 6 7 913 9 8 6 5                       6 7 91313 9 7 6 5
                6                           6
          91212 8 6                         6 7 912 9
      6 6 7         6 5                      6               7 6 6
          11 9 7                            711
          8 912     6 5 6     6    6 6 6 6     6       12 8
        6 7      11 8   7   7   8   7   7  11           7 6
              915  11 9   9    9   9  11       9
            7 8      13  12     13  12     13       7
            6        8 9  10     10  10     8        6
                 7       8     8     8     7
              6        7     7     7     6
                     6             6
                          5
```

Figure 2-8b. Apparent Viscosity.

```
                    27      29      27
                        23  24  23
              24          15  17  15      24
                  20      10  10  10    1420
          19          8   3   4   3   8      1519
              11       1  410  10   4   6 1   611
        14        2      12  15   15  15  12    4 2         14
          12          9  23  28   28  28292320 9       912
              7 5     2733  32   3232  33  27   1 5 7
      7          0 2 1                    1 1 0               7
        6 5 4 3 2                             2 4 5 6
              1 0 0 0                       0 0 1

    0 0 0 0 0 0 0 0 0                       0 0 0 0 0 0 0 0 0
                  0                           0
            4 3 2 0 0                         0 0 1 2 4
      7 6 5         1 1                       1            5 6 7
              2 0 2                             1 2
            9 7 5  202728  28   27322828  20     5 9
        1312       4 9  17  22   15  22  17   4          1213
            6 2     612  11   10  11   6      6
            1511      1   5     3   5     1      15
            19      13 7   8    10   8    13        19
                 19      15    17   15    19
                23        22   25   22   23
                      26          26
                          29
```

Figure 2-8c. Stress "AppVisc × dU(y,x)/dx".

```
                7         0    7
              6         0    6
      14      5         0    5      14
        12    4         0    4      912
  19      7  3         0    3      7    1519
    11    5  2 1       0    2      1 5   611
24      2    0 0       0    0      0   4 2        24
  20      9  2   0       0    0 1 220 9      1420
    8 1    27 1  0       0 0    1   27    6 1 8
27      122333                 291812              27
 231510 3 4                                    4101523
    10152832                     282210

29241710 410152832               282210 4 410172429
        32                       32
  10 3 41528                     282210 410
272315     2933                  29          152327
    61223                        18 6
  14 8 1  2027 1    0     2 0 0 1  20          114
 2420      5 9   2    0     3  0    2 5         2024
    6 1    1 1   0     2  0    1     6
   1511       4   1     1  1     4    15
   19       8 6   2     0  2     8       19
         11       4     0    4   11
       13       5     0    5   13
              6           6
              0
```

Figure 2-8d. Stress "AppVisc × dU(y,x)/dy".

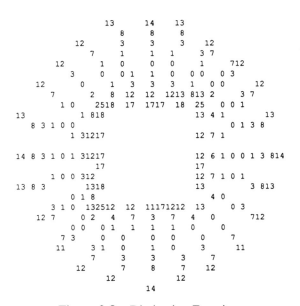

Figure 2-8e. Dissipation Function.

The comprehensive results computed for Examples 1 to 4 summarize the first suite of computations undertaken, where fluid model, properties and pressure gradient were fixed throughout, with only the annular geometry changing from run to run. The total flow rate trends and orders of magnitude for the physical quantities predicted are consistent with known empirical observation.

Next we describe results obtained for a "small diameter hole" and a "large diameter hole", using both a power law and a Bingham plastic model. These calculations were used in planning a horizontal well, using an experimental drilling fluid developed by a mud company. Since the exact rheology was open to question, two fluid models were used to bracket the performance of the mud insofar as hole cleaning was concerned. For further developments on hole cleaning, the reader should refer to Chapter 5 which deals with applications.

In the first set of calculations (Examples 5 and 6), the annular flow in the "small diameter hole" is evaluated for a Bingham plastic and for a power law fluid, assuming a volume flow rate of 500 gpm. The computer analysis iterates to find the appropriate pressure gradient, and terminates once the sought rate falls within 1% of the target. In the second set (Examples 7 and 8), similar calculations are pursued for the "large diameter hole". In the fifth and final simulation, a cuttings bed is added to the floor of the annulus. The results importantly show that cuttings beds do affect effective rheological properties, and that they ought to be included in routine operations planning.

Example 5:   Small Hole - Bingham Plastic Model

We will consider a pipe radius of 2.50 inches, and a hole radius of 4.25 inches; the pipe is displaced halfway down by 0.90 inches. The total volume flow rate is assumed to be 500 gpm. We will determine the pressure gradient required to support this flow rate and calculate detailed flow properties. Again, these methods do not involve any geometric approximation.

We solve this problem first assuming a Bingham plastic model, and next for a power law fluid. When the appropriate geometric parameters are entered, the computer model automatically generates a boundary conforming mesh which provides high resolution where physical gradients are large. This mesh is shown in Figure 2-9a.

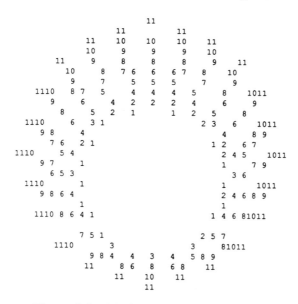

Figure 2-9a. Mesh System for Small Hole.

Again the actual coordinate lines are visually constructed by drawing ovals through lines of constant elevations, and then constructing orthogonals through these curves. We now take a Bingham plastic with a plastic viscosity of 25 cp, a yield stress of 0.00139 psi, and a target flow rate of 500 gpm.

The computer model tests different pressure gradients using a half-interval method, calculates the corresponding flow rates, and continues until the target rate of 500 gpm is met within 1%. These intermediate results also provide useful "pressure gradient vs flow rate" information for field applications. Linearity is expected only for Newtonian flows; in this particular run, the variation is "weakly nonlinear". From the computer output, we have specifically

O   Axial pressure gradient of .1500E-01 psi/ft
yields volume flow rate of .2648E+03 gal/min.

O   Axial pressure gradient of .2000E-01 psi/ft
yields volume flow rate of .3443E+03 gal/min.

O   Axial pressure gradient of .2500E-01 psi/ft
yields volume flow rate of .4084E+03 gal/min.

O   Axial pressure gradient of .3000E-01 psi/ft
yields volume flow rate of .4728E+03 gal/min.

and so on, until,

    O   Axial pressure gradient of .3187E-01 psi/ft
           yields volume flow rate of .4980E+03 gal/min.

           Pressure gradient found iteratively, .3187E-01 psi/ft,
           yielding .4980E+03 gal/min vs target .5000E+03 gal/min.

At this point, the velocity solution as it depends on annular position is obtained as shown in Figure 2-9b, and numerical results are written to output files for printing. Note how the no-slip condition is identically satisfied at all solid surfaces. Also, a plug flow regime having a speed of 65 in/sec was determined to occupy most of the annular space.

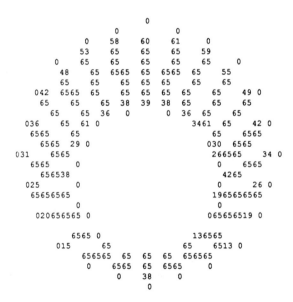

Figure 2-9b. Annular Velocity, Bingham Plug Flow in Small Hole.

```
                              L
                      L       T       L
                 L    T   T   T   T   L
                 T    T   T   T   T   T
            L    T    T   T   T   T       L
         T       T  T T   T  T T  T     T
         T       T   T    T   T   T     T
     L T   T T   T   T     T   T   T       T       T L
       T   T   T   T   T   T   T   T     T       T
         T     T   T   L      L   T   T       T
     L T     T   T L               T T   T       T L
       T T     T                        T       T T
         T T     T L                   L T     T T
     L T     T T                        T T T       T L
       T T     L                        L       T T
         T T T                               T T
     L L       L                        L           T L
       T T T T                           L T T T T
               L                         L
       L L T T T L                   L T T T L L

           T T L                     L T T
         L L       T               T       T L L
           T T T   T   T   T   T T T
           L     T T   T   T T     L
               L     T     L
                     L
```

Figure 2-9c. Laminar and Turbulent Flow Regimes.

```
                              26
                      25              25
              23      17      18  18      22
              15      10      11  11      15
        18     9       5       5   5       9       18
          12        4     1 2   3   3 0     4      13
           7        0       7   7   7       0       7
     12 8   3 0   6      11  12  12   7       3       812
        4       2    10  20  20  20  11       3       5
         1       5  18  18        18  18   6       1
      6 3      2  1313                1413    3      3 6
        1 0       7                        7       0 1
          1 2  10 9                      910     2 1
     0 0      3 4                          5 4 3       1 0
       1 1       5                         5       2 1
         1 2 1                               1 2
     7 5         0                         0         5 6
       3 3 1 0                           4 1 1 4 5
               4                         4
     13 9 4 0 3 8                        8 3 1 5 912

           2 317                     12 1 3
        1912       10              9       61217
             8 4 7   8   10    6    2 6 9
             15      6 0    5    0 6      22
                 25         7    13
                            4
```

Figure 2-9d. Stress "AppVisc × dU(y,x)/dx".

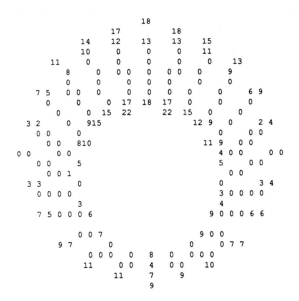

Figure 2-9e. Shear Rate dU(y,x)/dx.

Stability analyses are extremely difficult to carry out for eccentric flows, and no claim is made to have solved the problem in any form. However, a facility was created to provide a display of local Reynolds number, based on borehole diameter, local velocity and apparent viscosity. These values are then checked against a critical Reynolds number R_C entered by the user to produce a "laminar versus turbulent flow map". In the following, a mud weight 1.5 times that of water and a R_C of 2000 was assumed. In Figure 2-9c, the program displays "L" for laminar and "T" for turbulent flow.

Also, average Reynolds numbers are obtained and printed in output files, for example,

Average Reynolds number, bottom half annulus = .3770E+04
Average Reynolds number, entire annulus = .3985E+04

Two components of viscous stress are again computed. The distribution in the vertical direction, which is particularly important to cuttings bed removal, is shown in Figure 2-9d. The shear rate corresponding to this stress distribution is also obtained as a part of the exact solution; its computed values (in reciprocal seconds) within the annular geometry are shown in Figure 2-9e. Note how "zeros" correctly indicate zero shear in plug dominated regions.

The annular flow model also provides areal averages of all computed quantities, for the "bottom half" of the annulus where drilled cuttings are known to settle, and for the entire annulus. The former are useful in assessing potential dangers in cuttings removal (see Chapter 5 for details). Both averages are summarized in Table 2-11.

Table 2-11
Example 5: Summary, Average Quantities

TABULATION OF CALCULATED AVERAGE QUANTITIES
Area weighted means of absolute values taken over
BOTTOM HALF of annular cross-section ...
O Average annular velocity = .4445E+02 in/sec
O Average stress, AppVis x dU/dx, = .8316E-03 psi
O Average stress, AppVis x dU/dy, = .7333E-03 psi
O Average dissipation = .1045E+00 lbf/(sec sq in)
O Average shear rate dU/dx = .3384E+02 1/sec
O Average shear rate dU/dy = .3123E+02 1/sec

TABULATION OF CALCULATED AVERAGE QUANTITIES
Area weighted means of absolute values taken over
ENTIRE annular (y,x) cross-section ...
O Average annular velocity = .4662E+02 in/sec
O Average stress, AppVis x dU/dx, = .8169E-03 psi
O Average stress, AppVis x dU/dy, = .8032E-03 psi
O Average dissipation = .1324E+00 lbf/(sec sq in)
O Average shear rate dU/dx = .4047E+02 1/sec
O Average shear rate dU/dy = .4038E+02 1/sec

Finally, plots of all computed quantities in the two vertical planes above and below the drillpipe, using convenient built-in graphical software, are always available for all runs. Self-explanatory plots are shown in Figures 2-9f and 2-9g for velocities and viscous stresses. Note that the values of stress are not entirely zero inside the plug region; but they are less than yield thresholds as required.

VERTICAL SYMMETRY PLANE ABOVE DRILL PIPE
Axial velocity distribution (in/sec):

```
   X                    0
                        _____
  1.00    .0000E+00   |
  1.38    .6057E+02   |                      *
  1.72    .6591E+02   |                        *
  2.04    .6591E+02   |                        *
  2.32    .6591E+02   |                        *
  2.59    .6591E+02   |                        *
  2.84    .6591E+02   |                        *
  3.06    .6591E+02   |                        *
  3.27    .6591E+02   |                        *
  3.47    .3920E+02   |              *
  3.65    .0000E+00   |
```

VERTICAL SYMMETRY PLANE BELOW DRILL PIPE
Axial velocity distribution (in/sec):

```
   X                    0
                        _____
  8.65    .0000E+00   |
  8.70    .9162E+01   | *
  8.76    .6591E+02   |                        *
  8.83    .6591E+02   |                        *
  8.90    .6591E+02   |                        *
  8.98    .6591E+02   |                        *
  9.06    .6591E+02   |                        *
  9.16    .6591E+02   |                        *
  9.26    .6591E+02   |                        *
  9.37    .3833E+02   |                *
  9.50    .0000E+00   |
```

Figure 2-9f.   Vertical Plane Velocity Plots

VERTICAL SYMMETRY PLANE ABOVE DRILL PIPE
Viscous stress, AppVis x dU/dx (psi):

X			0
1.00	-.2671E-02		
1.38	-.1882E-02	*	|
1.72	-.1136E-02	*	|
2.04	-.5755E-03	*	|
2.32	-.1058E-03		*|
2.59	.3245E-03		|*
2.84	.7604E-03		| *
3.06	.1246E-02		| *
3.27	.1881E-02		| *
3.47	.2071E-02		| *
3.65	.1960E-02		| *

VERTICAL SYMMETRY PLANE BELOW DRILL PIPE
Viscous stress, AppVis x dU/dx (psi):

X			0
8.65	-.2312E-02	*	|
8.70	-.1901E-02	*	|
8.76	-.1446E-02	*	|
8.83	-.8806E-03	*	|
8.90	-.4380E-03	*	|
8.98	-.1057E-04		*|
9.06	.4153E-03		|*
9.16	.8532E-03		| *
9.26	.1169E-02		| *
9.37	.2245E-02		| *
9.50	.3567E-02		| *

Figure 2-9g. Vertical Plane Viscous Stress Plots

Example 6:   Small Hole - Power Law Fluid

We now repeat Example 5, but assume a power law fluid model. The power law index n and the consistency factor k can be inputted directly as in Examples 1 - 4. However, the program also allows users to input known Fann dial readings, from which n and k values are internally calculated. In the present case we choose the latter option. Intermediate pressure and flow rate results are tabulated below.

POWER LAW FLOW OPTION SELECTED

1st Fann dial reading of .1500E+02 with corresponding
rpm of .1300E+02 assumed.

2nd Fann dial reading of .2000E+02 with corresponding
rpm of .5000E+02 assumed. We calculate "n" and "k".

Power law fluid assumed, with exponent "n" equal
to .2136E+00 and consistency factor of .5376E-03
lbf sec^n/sq in.

Target flow rate of .5000E+03 gal/min specified.

The program iterates on pressure gradient to match the flow rate ...

O   Axial pressure gradient of .5000E-02 psi/ft
       yields volume flow rate of .5250E-02 gal/min.

O   Axial pressure gradient of .1000E-01 psi/ft
       yields volume flow rate of .1349E+00 gal/min.

O   Axial pressure gradient of .1500E-01 psi/ft
       yields volume flow rate of .9035E+00 gal/min.
       .
       .

O   Axial pressure gradient of .4500E-01 psi/ft
       yields volume flow rate of .1546E+03 gal/min.

O   Axial pressure gradient of .5000E-01 psi/ft
       yields volume flow rate of .2533E+03 gal/min.

until the following converged result is obtained,

O   Axial pressure gradient of .5781E-01 psi/ft
    yields volume flow rate of .5003E+03 gal/min.

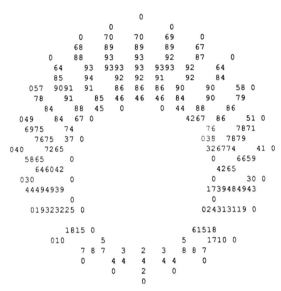

Figure 2-10a. Annular Velocity, Power Law Flow in Small Hole.

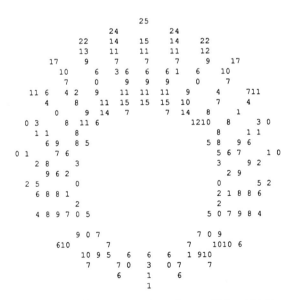

Figure 2-10b. Stress "AppVisc × dU(y,x)/dx".

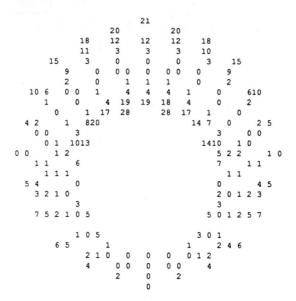

Figure 2-10c. Shear Rate dU(y,x)/dx.

Figure 2-11a. Mesh System for Large Hole.

The large pressure gradients here and in Example 5 are in rough agreement with internal laboratory data obtained by two different oil companies for this experimental mud.

Because power law models exclude the possibility of plug flow, the velocity distribution calculated is somewhat different. In Figure 2-10a, note how no-slip conditions are again satisfied, and how velocity maximums are correctly obtained near the "center" of the annulus. Figures 2-10b and 2-10c show computed stresses and shear rates.

From Table 2-12 we observe that the "bottom half" average velocities are one-third of those for the entire annulus. Plots for exact annular velocities above and below the drill pipe in Figure 2-10d show even larger contrasts.

Table 2-12
Example 6: Summary, Average Quantities

TABULATION OF CALCULATED AVERAGE QUANTITIES
Area weighted means of absolute values taken over
BOTTOM HALF of annular cross-section ...
O Average annular velocity = .1360E+02 in/sec
O Average apparent viscosity = .6028E-04 lbf sec/sq in
O Average stress, AppVis x dU/dx, = .6395E-03 psi
O Average stress, AppVis x dU/dy, = .6879E-03 psi
O Average dissipation = .4688E-01 lbf/(sec sq in)
O Average shear rate dU/dx = .2139E+02 1/sec
O Average shear rate dU/dy = .3168E+02 1/sec
O Average Stokes product = .6083E-03 lbf/in

TABULATION OF CALCULATED AVERAGE QUANTITIES
Area weighted means of absolute values taken over
ENTIRE annular (y,x) cross-section ...
O Average annular velocity = .3841E+02 in/sec
O Average apparent viscosity = .6230E-04 lbf sec/sq in
O Average stress, AppVis x dU/dx, = .7269E-03 psi
O Average stress, AppVis x dU/dy, = .7239E-03 psi
O Average dissipation = .9535E-01 lbf/(sec sq in)
O Average shear rate dU/dx = .4191E+02 1/sec
O Average shear rate dU/dy = .4855E+02 1/sec
O Average Stokes product = .3266E-02 lbf/in

VERTICAL SYMMETRY PLANE ABOVE DRILL PIPE
Axial velocity distribution (in/sec):

```
     X                    0

   1.00    .0000E+00   |
   1.38    .7068E+02   |                  *
   1.72    .8983E+02   |                     *
   2.04    .9341E+02   |                     *
   2.32    .9361E+02   |                      *
   2.59    .9353E+02   |                     *
   2.84    .9219E+02   |                     *
   3.06    .8696E+02   |                  *
   3.27    .7368E+02   |              *
   3.47    .4686E+02   |        *
   3.65    .0000E+00   |
```

VERTICAL SYMMETRY PLANE BELOW DRILL PIPE
Axial velocity distribution (in/sec):

```
     X                    0

   8.65    .0000E+00   |
   8.70    .1642E+01   |        *
   8.76    .2797E+01   |             *
   8.83    .3560E+01   |                *
   8.90    .4019E+01   |                   *
   8.98    .4244E+01   |                   *
   9.06    .4265E+01   |                    *
   9.16    .4036E+01   |                   *
   9.26    .3419E+01   |                *
   9.37    .2188E+01   |           *
   9.50    .0000E+00   |
```

Figure 2-10d.   Vertical Plane Velocity Plots

Example 7:   Large Hole - Bingham Plastic

We next consider a 2.5 inch radius drillpipe residing in a 6.1 inch radius borehole. The pipe is displaced halfway down by 1.80 inches. Again, the computer code produces a boundary conforming grid system for exact calculations. This mesh is shown in Figure 2-11a.

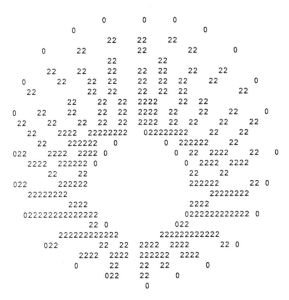

Figure 2-11b. Annular Velocity, Bingham Plug Flow in Large Hole.

We will first consider a Bingham plastic fluid. The plastic viscosity is taken to be 25 cp, the yield stress as 0.00139 psi, and the target flow rate again 500 gpm. Iterating on pressure gradient to match flow rate, we find the following "pressure gradient versus flow rate" signature.

O   Axial pressure gradient of .5000E-02 psi/ft
    yields volume flow rate of .7026E+03 gal/min.

O   Axial pressure gradient of .3750E-02 psi/ft
    yields volume flow rate of .5270E+03 gal/min.

O   Axial pressure gradient of .3125E-02 psi/ft
    yields volume flow rate of .4391E+03 gal/min.

O   Axial pressure gradient of .3555E-02 psi/ft
yields volume flow rate of .4995E+03 gal/min.

Pressure gradient found iteratively, .3555E-02 psi/ft,
yielding .4995E+03 gal/min vs target .5000E+03 gal/min.

Since the annular space is large compared with the "small diameter" run, the computed viscous stresses are smaller. In this particular calculation, the results indicate that they are well below yield. Hence, the entire annular flow moves as a solid plug at a constant velocity of 22 in/sec as shown in Figure 2-11b. At solid surfaces, the velocity profile rapidly adjusts to "0" to satisfy no-slip conditions, thus giving an average speed of approximately 18 in/sec.

Table 2-13
Example 7:   Summary, Average Quantities

TABULATION OF CALCULATED AVERAGE QUANTITIES
Area weighted means of absolute values taken over
BOTTOM HALF of annular cross-section ...
O Average annular velocity = .1870E+02 in/sec
O Average stress, AppVis x dU/dx, = .1966E-03 psi
O Average stress, AppVis x dU/dy, = .1656E-03 psi
O Average dissipation = .4930E-02 lbf/(sec sq in)
O Average shear rate dU/dx = .6637E+01 1/sec
O Average shear rate dU/dy = .4466E+01 1/sec

TABULATION OF CALCULATED AVERAGE QUANTITIES
Area weighted means of absolute values taken over
ENTIRE annular (y,x) cross-section ...
O Average annular velocity = .1870E+02 in/sec
O Average stress, AppVis x dU/dx, = .1771E-03 psi
O Average stress, AppVis x dU/dy, = .1716E-03 psi
O Average dissipation = .4724E-02 lbf/(sec sq in)
O Average shear rate dU/dx = .5450E+01 1/sec
O Average shear rate dU/dy = .4794E+01 1/sec

Example 8: Large Hole - Power Law Fluid

Now we repeat Example 7, assuming instead a power law fluid. The results for n and k are summarized below, as are the computed quantities.

POWER LAW FLOW OPTION SELECTED

1st Fann dial reading of .1500E+02 with corresponding rpm of .1300E+02 assumed.

2nd Fann dial reading of .2000E+02 with corresponding rpm of .5000E+02 assumed. We calculate "n" and "k".

Power law fluid assumed, with exponent "n" equal to .2136E+00 and consistency factor of .5376E-03 lbf sec^n/sq in.

Target flow rate of .5000E+03 gal/min specified.

Iterating on pressure gradient to match flow rate ...

  O   Axial pressure gradient of .5000E-02 psi/ft
        yields volume flow rate of .6528E+00 gal/min.

  O   Axial pressure gradient of .1000E-01 psi/ft
        yields volume flow rate of .1593E+02 gal/min.

and so on, until,

  O   Axial pressure gradient of .2094E-01 psi/ft
        yields volume flow rate of .5039E+03 gal/min.

The computed velocity field shown in Figure 2-12a is somewhat more interesting than the plug flow obtained in Figure 2-11b. Other calculated quantities are given in Figures 2-12b to 2-12g without explanation. Relevant areal averages are listed in Table 2-14.

```
               0         0    0
          0                        0
              28   28   28
     0     27          32        27        0
              32        32
       26   32   32    32  32    32   26
   0     31    32  32   32  32  32    31        0
    24        32  32    32  32  32           24
         32    32  31  3131    32  32
   0   30    32    31  29 2929  31    32    30    0
   22   31    32  28  23 2323  28  32    31    22
    28    3131  27231414  014142327    31    28
     29    292622    0      0 222629    30
   018    3029  2112 0      0  21  2930    19    0
    2527  272418 0          0  2427  2725
        27    10              11    27
   015    2421 9              92124        15 0
    20222221                  21222221
          1712                  717
    0111517171512 5      0121517171511 0
            6 0                  0 6
      101111  8  4 2        2  4  8111110
    0 7          7   3    3 3    6 7        7 0
            4 7    4 4    4 4 4    7 4
              0      3    3 3    3        0
              0 2      2    0
                       0
```

Figure 2-12a. Annular Velocity, Power Law Flow in Large Hole.

Table 2-14

Example 8:   Summary, Average Quantities

TABULATION OF CALCULATED AVERAGE QUANTITIES
Area weighted means of absolute values taken over
BOTTOM HALF of annular cross-section ...
O Average annular velocity = .7696E+01 in/sec
O Average apparent viscosity = .1198E-03 lbf sec/sq in
O Average stress, AppVis x dU/dx, = .5251E-03 psi
O Average stress, AppVis x dU/dy, = .5130E-03 psi
O Average dissipation = .1034E-01 lbf/(sec sq in)
O Average shear rate dU/dx = .6954E+01 1/sec
O Average shear rate dU/dy = .8676E+01 1/sec
O Average Stokes product = .9454E-03 lbf/in

TABULATION OF CALCULATED AVERAGE QUANTITIES
Area weighted means of absolute values taken over
ENTIRE annular (y,x) cross-section ...
O Average annular velocity = .1527E+02 in/sec
O Average apparent viscosity = .2074E-03 lbf sec/sq in
O Average stress, AppVis x dU/dx, = .5562E-03 psi
O Average stress, AppVis x dU/dy, = .5272E-03 psi
O Average dissipation = .1530E-01 lbf/(sec sq in)
O Average shear rate dU/dx = .9280E+01 1/sec
O Average shear rate dU/dy = .1056E+02 1/sec
O Average Stokes product = .5144E-02 lbf/in

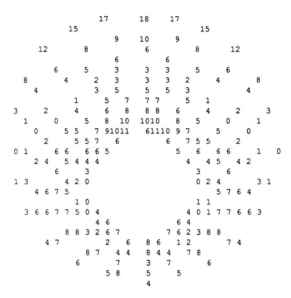

Figure 2-12b. Stress "AppVisc × dU(y,x)/dx".

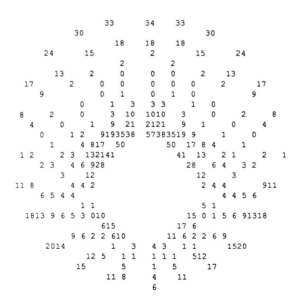

Figure 2-12c. Shear Rate dU(y,x)/dx.

VERTICAL SYMMETRY PLANE ABOVE DRILL PIPE
Axial velocity distribution (in/sec):

```
     X                    0
                          _____
   1.00    .0000E+00     |
   1.95    .2859E+02     |                       *
   2.76    .3260E+02     |                         *
   3.45    .3272E+02     |                          *
   4.05    .3271E+02     |                         *
   4.58    .3258E+02     |                         *
   5.03    .3175E+02     |                         *
   5.44    .2928E+02     |                      *
   5.80    .2399E+02     |                 *
   6.12    .1463E+02     |          *
   6.40    .0000E+00     |
```

VERTICAL SYMMETRY PLANE BELOW DRILL PIPE
Axial velocity distribution (in/sec):

```
     X                    0
                          _____
  11.40    .0000E+00     |
  11.51    .1649E+01     |     *
  11.64    .2890E+01     |          *
  11.77    .3762E+01     |             *
  11.92    .4313E+01     |               *
  12.09    .4595E+01     |                *
  12.27    .4640E+01     |                 *
  12.47    .4404E+01     |                *
  12.69    .3741E+01     |             *
  12.93    .2403E+01     |         *
  13.20    .0000E+00     |
```

Figure 2-12d.   Vertical Plane Plots

VERTICAL SYMMETRY PLANE ABOVE DRILL PIPE
Apparent viscosity distribution (lbf sec/sq in):

```
     X                    0

  1.00    .5410E-04    |
  1.95    .5410E-04    |
  2.76    .2425E-03    *
  3.45    .3657E-02    |                      *
  4.05    .2792E-02    |                *
  4.58    .5468E-03    | *
  5.03    .1873E-03    *
  5.44    .8667E-04    |
  5.80    .4789E-04    |
  6.12    .2969E-04    |
  6.40    .1149E-04    |
```

VERTICAL SYMMETRY PLANE BELOW DRILL PIPE
Apparent viscosity distribution (lbf sec/sq in):

```
     X                    0

 11.40    .6464E-04    | *
 11.51    .1046E-03    |  *
 11.64    .1446E-03    |    *
 11.77    .2176E-03    |       *
 11.92    .3630E-03    |          *
 12.09    .5925E-03    |               *
 12.27    .5426E-03    |             *
 12.47    .3238E-03    |         *
 12.69    .1997E-03    |      *
 12.93    .1332E-03    |   *
 13.20    .6672E-04    | *
```

Figure 2-12e. Vertical Plane Plots

VERTICAL SYMMETRY PLANE ABOVE DRILL PIPE
Viscous stress, AppVis x dU/dx (psi):

```
     X                         0
         _____
1.00    -.1857E-02                  |
1.95    -.1003E-02          *       |
2.76    -.6673E-03             *    |
3.45    -.3194E-03                * |
4.05     .3437E-03                  | *
4.58     .5351E-03                  |   *
5.03     .7158E-03                  |    *
5.44     .8824E-03                  |      *
5.80     .1037E-02                  |       *
6.12     .1180E-02                  |         *
6.40     .6649E-03                  |    *
```

VERTICAL SYMMETRY PLANE BELOW DRILL PIPE
Viscous stress, AppVis x dU/dx (psi):

```
     X                         0
          _____
11.40    -.6990E-03         *       |
11.51    -.8801E-03                 |
11.64    -.8693E-03                 |
11.77    -.8743E-03                 |
11.92    -.9128E-03                 |
12.09    -.8764E-03                 |
12.27    -.3433E-03            *     |
12.47     .1435E-03                 | *
12.69     .3946E-03                 |    *
12.93     .5310E-03                 |      *
13.20     .4001E-03                 |    *
```

Figure 2-12f.   Vertical Plane Plots

VERTICAL SYMMETRY PLANE ABOVE DRILL PIPE
Shear rate dU/dx (1/sec):

```
   X                                    0
        _____
 1.00    -.3433E+02        *      |
 1.95    -.1854E+02            *  |
 2.76    -.2752E+01              *|
 3.45    -.8735E-01              *|
 4.05     .1231E+00              |
 4.58     .9787E+00              |
 5.03     .3821E+01              |
 5.44     .1018E+02             | *
 5.80     .2165E+02             |   *
 6.12     .3976E+02             |       *
 6.40     .5787E+02             |           *
```

VERTICAL SYMMETRY PLANE BELOW DRILL PIPE
Shear rate dU/dx (1/sec):

```
   X                                    0
        _____
11.40    -.1081E+02                    |
11.51    -.8412E+01       *            |
11.64    -.6012E+01          *         |
11.77    -.4018E+01             *      |
11.92    -.2515E+01               *    |
12.09    -.1479E+01                 *  |
12.27    -.6328E+00                  *|
12.47     .4433E+00                   |
12.69     .1976E+01                  | *
12.93     .3986E+01                  |   *
13.20     .5997E+01                  |     *
```

Figure 2-12g.   Vertical Plane Plots

Example 9:   Large Hole with Cuttings Bed

Finally, the geometry of Example 8 was re-evaluated to examine the effect of a flat cuttings bed, using the pressure gradient obtained in the above run. The boundary conforming mesh generated for this annulus resolves the flowfield at the bed in sufficient detail, and provides as many "radial" meshes at the bottom as there on top. The computed mesh is shown in Figure 2-13a. Figures 2-13b to 2-13f, and Table 2-15, summarize the self-explanatory simulation results. To display plotted precisely, commercial plotting packages may be used. Some partial results follow.

Table 2-15
Example 9:   Summary, Average Quantities

TABULATION OF CALCULATED AVERAGE QUANTITIES
Area weighted means of absolute values taken over
BOTTOM HALF of annular cross-section ...
O Average annular velocity = .6708E+01 in/sec
O Average apparent viscosity = .1258E-03 lbf sec/sq in
O Average stress, AppVis x dU/dx, = .4717E-03 psi
O Average stress, AppVis x dU/dy, = .5651E-03 psi
O Average dissipation = .9985E-02 lbf/(sec sq in)
O Average shear rate dU/dx = .6188E+01 1/sec
O Average shear rate dU/dy = .8557E+01 1/sec
O Average Stokes product = .6989E-03 lbf/in

TABULATION OF CALCULATED AVERAGE QUANTITIES
Area weighted means of absolute values taken over
ENTIRE annular (y,x) cross-section ...
O Average annular velocity = .1483E+02 in/sec
O Average apparent viscosity = .2098E-03 lbf sec/sq in
O Average stress, AppVis x dU/dx, = .5320E-03 psi
O Average stress, AppVis x dU/dy, = .5513E-03 psi
O Average dissipation = .1517E-01 lbf/(sec sq in)
O Average shear rate dU/dx = .8943E+01 1/sec
O Average shear rate dU/dy = .1052E+02 1/sec
O Average Stokes product = .5027E-02 lbf/in

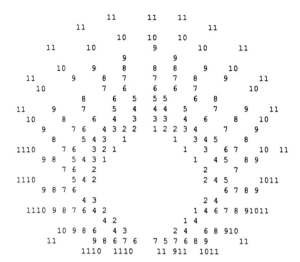

```
              11        11   11
        11                        11
                 10   10   10
     11       10          9          10      11
                   9           9
        10       9     8    8  8    9    10
    11       9     8  7    7  7   8     9       11
     10             7  6    6  6  7            10
               8     6 5  5 5     6  8
    11     9     7     5  4  4 4    5     7     9    11
     10     8     6   4  3  3 3   4  6     8     10
       9     7 6   4 3 2 2   1 2 2 3 4     7        9
         8     5 4 3   1     1   3 4 5     8
    1110     7 6   3 2 1        1   3    6 7    10 11
      9 8    5 4 3 1          1   4 5    8 9
        7 6   2               2      7
    1110     5 4 2           2 4 5        1011
      9 8 7 6                6 7 8 9
          4 3               2 4
    1110 9 8 7 6 4 2        1 4 6 7 8 91011
            4 2            1 4
      10 9 8 6    4 3      2 4   6 8 910
    11        9 8 6 7 6  7 5 7 6 8 9        11
          1110    1110     11 911  1011
```

Figure 2-13a. Mesh System, Large Hole with Cuttings Bed.

```
                 0        0    0
             0                     0
                   28    28   28
          0       27          32          27       0
                      32          32
            26    32     32    32 32    32    26
       0     31      32  32    32 32 32    31        0
        24            32 32    32 32 32              24
                 32      32 31 3131    32  32
     0    30      32     31 29 2929 31    32     30     0
      22    31      32 28 23 2423 28 32    31    22
       28      3131  27231414 014142328    31    28
         30      292622  0     0 222629    30
     019    3029  2112 0       0  21 2930   19    0
       2527  272418 0            0 2427  2725
         2726  10               10    27
     015      2420 9              92124        15 0
       21222221                  21222221
            1512                   715
     0111517171512 4         0121517171511 0
            7 2                0 7
          7101110   3 2        1 3  101110 7
        0        5 5 4 1 1  1 1 1 4 5 5         0
            0 4    0 0       0   1 0    4 0
```

Figure 2-13b. Annular Velocity.

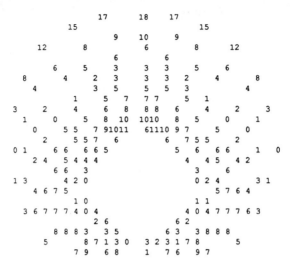

Figure 2-13c. Stress "AppVisc × dU(y,x)/dx".

Figure 2-13d. Shear Rate dU(y,x)/dx.

POWER LAW FLOW OPTION SELECTED

1st Fann dial reading of .1500E+02 with corresponding
rpm of .1300E+02 assumed.

2nd Fann dial reading of .2000E+02 with corresponding
rpm of .5000E+02 assumed.   We calculate "n" and "k"
(continued next page).

VERTICAL SYMMETRY PLANE ABOVE DRILL PIPE
Axial velocity distribution (in/sec):

```
     X                    0
                     _____
   1.00    .0000E+00  |
   1.95    .2863E+02  |                      *
   2.76    .3265E+02  |                        *
   3.45    .3277E+02  |                         *
   4.05    .3277E+02  |                        *
   4.58    .3263E+02  |                        *
   5.03    .3180E+02  |                        *
   5.44    .2932E+02  |                     *
   5.80    .2402E+02  |                *
   6.12    .1464E+02  |           *
   6.40    .0000E+00  |
```

VERTICAL SYMMETRY PLANE BELOW DRILL PIPE
Axial velocity distribution (in/sec):

```
     X                    0
                     _____
  11.40    .0000E+00  |
  11.44    .2826E+00  |  *
  11.49    .5483E+00  |     *
  11.55    .7916E+00  |        *
  11.60    .1006E+01  |           *
  11.67    .1183E+01  |             *
  11.74    .1306E+01  |               *
  11.82    .1346E+01  |                *
  11.90    .1242E+01  |              *
  12.00    .8765E+00  |        *
  12.10    .0000E+00  |
```

Figure 2-13e.   Vertical Plane Plots

Power law fluid assumed, with exponent "n" equal to .2136E+00 and consistency factor of .5376E-03 lbf $\sec^n$/sq in.

Axial pressure gradient assumed as .2094E-01 psi/ft.

VERTICAL SYMMETRY PLANE ABOVE DRILL PIPE
Apparent viscosity distribution (lbf sec/sq in):

X	0	
1.00	.5404E-04	I
1.95	.5404E-04	I
2.76	.2423E-03	*
3.45	.3657E-02	I *
4.05	.2781E-02	I *
4.58	.5451E-03	I *
5.03	.1869E-03	*
5.44	.8651E-04	I
5.80	.4782E-04	I
6.12	.2965E-04	I
6.40	.1149E-04	I

VERTICAL SYMMETRY PLANE BELOW DRILL PIPE
Apparent viscosity distribution (lbf sec/sq in):

X	0	
11.40	.2545E-03	I *
11.44	.2981E-03	I *
11.49	.3417E-03	I *
11.55	.3963E-03	I *
11.60	.4519E-03	I *
11.67	.4709E-03	I *
11.74	.4099E-03	I *
11.82	.3051E-03	I *
11.90	.2146E-03	I *
12.00	.1496E-03	I *
12.10	.8459E-04	I *

Figure 2-13f.   Vertical Plane Plots

We have presented in this chapter, a self-contained mathematical model describing the flow of Newtonian, power law, Bingham plastic and Herschel-Bulkley fluids through eccentric annular spaces. Again, the borehole contour need not be circular; in fact, it may be modified "point by point" to simulate cuttings beds and general wall deformations due to erosion and swelling (see Chapter 5). Likewise, the default "pipe contour" while circular, may be altered to model drill pipes and collars with or without stabilizers and centralizers.

A fast, second-order accurate, unconditionally stable finite difference scheme was used to solve the nonlinear governing PDEs. The formulation uses exact boundary conforming grid systems which eliminate the need for unrealistic simplifying assumptions about the annular geometry.

Computed quantities include detailed plots and tables for annular velocity, apparent viscosity, viscous stress, shear rate, Stokes product and dissipation function. To facilitate visual correlation with annular position, a portable, Fortran-based, ASCII text plotter was developed to overlay computed quantities directly on the annulus. Software displaying "vertical plane plots" was also written to enhance the interpretation of computed quantities.

Calculations for several non-Newtonian flows using a number of complicated annular geometries were performed. The results, which agree with generally accepted empirical observation, were computed in a stable manner in all cases. The cross-sectional displays show an unusual amount of information that is easily interpreted and understood by drilling and production engineers. Moreover, the computer algorithm is fast and requires minimal hardware and graphical software investment.

REFERENCES

---

1. Lapidus, L., and Pinder, G., *Numerical Solution of Partial Differential Equations in Science and Engineering,* New York: John Wiley & Sons, 1982.
2. Crochet, M.J., Davies, A.R., and Walters, K., *Numerical Simulation of Non-Newtonian Flow,* Amsterdam: Elsevier Science Publishers B.V., 1984.
3. Thompson, J.F., Warsi, Z.U.A., and Mastin, C.W., *Numerical Grid Generation,* New York: Elsevier Science Publishing, 1985.
4. Bird, R.B., Stewart, W.E., and Lightfoot, E.N., *Transport Phenomena,* New York: John Wiley & Sons, 1960.
5. Schlichting, H., *Boundary Layer Theory,* New York: McGraw-Hill, 1968.
6. Slattery, J.C., *Momentum, Energy, and Mass Transfer in Continua,* New York: Robert E. Krieger Publishing Company, 1981.
7. Streeter, V.L., *Handbook of Fluid Dynamics,* New York: McGraw-Hill, 1961.

3
Concentric, Rotating Annular Flow

Analytical solutions for the nonlinearly coupled axial and circumferential velocities, their deformation, stress and pressure fields, are obtained for the annular flow in an inclined borehole with a centered rotating drillstring or casing. The closed form solutions are used to derive formulas for volume flow rate, maximum borehole wall stress, apparent viscosity and other quantities as functions of "r".

The analysis is restricted to Newtonian and power law fluids. Our Newtonian results are *exact* solutions to the viscous Navier-Stokes equations. For power law fluids, the analytical results assume a *narrow annulus*, but reduce to the Newtonian solutions in the "n=1" limit. All solutions satisfy no-slip viscous boundary conditions at both the rotating drillstring and the borehole wall. The formulas are also explicit; they require no iteration and are easily programmed on pocket calculators. Extensive analytical and calculated results are given, which elucidate the physical differences between the two fluid types.

GENERAL GOVERNING EQUATIONS

The equations governing general fluid motion are available from many excellent textbooks on continuum mechanics (Schlichting, 1968; Slattery, 1981). We will cite these equations without proof. Let v_r, v_ϑ and v_z denote Eulerian fluid velocities, and F_r, F_ϑ and F_z the body forces in the r, ϑ and z directions, respectively. Here (r,ϑ,z) are standard circular cylindrical coordinates.

Also, let ρ be the *constant* fluid density and p be the pressure; and denote by S_{rr}, $S_{r\vartheta}$, $S_{\vartheta\vartheta}$, S_{rz}, $S_{\vartheta r}$, $S_{\vartheta z}$, S_{zr}, $S_{z\vartheta}$ and S_{zz} the nine elements of the general extra stress tensor $\underline{\underline{S}}$. If t is time, and ∂'s represent partial derivatives, the complete equations obtained from Newton's law and mass conservation are

Momentum equation in r:

$$\rho \, (\partial v_r/\partial t + v_r \, \partial v_r/\partial r + v_\vartheta/r \, \partial v_r/\partial\vartheta - v_\vartheta^2/r + v_z \, \partial v_r/\partial z) =$$
$$= F_r - \partial p/\partial r + 1/r \, \partial(rS_{rr})/\partial r + 1/r \, \partial(S_{r\vartheta})/\partial\vartheta$$
$$+ \partial(S_{rz})/\partial z - S_{\vartheta\vartheta}/r \qquad (3\text{-}1)$$

Momentum equation in ϑ:

$$\rho \, (\partial v_\vartheta/\partial t + v_r \, \partial v_\vartheta/\partial r + v_\vartheta/r \, \partial v_\vartheta/\partial\vartheta + v_r v_\vartheta/r + v_z \, \partial v_\vartheta/\partial z) =$$
$$= F_\vartheta - 1/r \, \partial p/\partial\vartheta + 1/r^2 \, \partial(r^2 S_{\vartheta r})/\partial r + 1/r \, \partial(S_{\vartheta\vartheta})/\partial\vartheta$$
$$+ \partial(S_{\vartheta z})/\partial z \qquad (3\text{-}2)$$

Momentum equation in z:

$$\rho \, (\partial v_z/\partial t + v_r \, \partial v_z/\partial r + v_\vartheta/r \, \partial v_z/\partial\vartheta + v_z \, \partial v_z/\partial z) =$$
$$= F_z - \partial p/\partial z + 1/r \, \partial(rS_{zr})/\partial r + 1/r \, \partial(S_{z\vartheta})/\partial\vartheta$$
$$+ \partial(S_{zz})/\partial z \qquad (3\text{-}3)$$

Continuity equation:

$$1/r \, \partial(rv_r)/\partial r + 1/r \, \partial v_\vartheta/\partial\vartheta + \partial v_z/\partial z = 0 \qquad (3\text{-}4)$$

These equations apply to all Newtonian and non-Newtonian fluids. In continuum mechanics, the most common class of empirical models for incompressible fluids assumes that $\underline{\underline{S}}$ can be related to the rate of deformation tensor $\underline{\underline{D}}$ by a relationship of the form

$$\underline{\underline{S}} = 2 \, N(\Gamma) \, \underline{\underline{D}} \qquad (3\text{-}5)$$

where the elements of $\underline{\underline{D}}$ are

$$D_{rr} = \partial v_r/\partial r \qquad (3\text{-}6)$$

$$D_{\vartheta\vartheta} = 1/r \, \partial v_\vartheta/\partial\vartheta + v_r/r \qquad (3\text{-}7)$$

$$D_{zz} = \partial v_z/\partial z \qquad (3\text{-}8)$$

$$D_{r\vartheta} = D_{\vartheta r} = [r \, \partial(v_\vartheta/r)/\partial r + 1/r \, \partial v_r/\partial\vartheta]/2 \qquad (3\text{-}9)$$

$$D_{rz} = D_{zr} = [\partial v_r/\partial z + \partial v_z/\partial r]/2 \qquad (3\text{-}10)$$

$$D_{\vartheta z} = D_{z\vartheta} = [\partial v_\vartheta/\partial z + 1/r \, \partial v_z/\partial\vartheta]/2 \qquad (3\text{-}11)$$

assuming isotropic flow. In Equation 3-5, $N(\Gamma)$ is the well known "apparent viscosity function" satisfying

$$N(\Gamma) > 0 \qquad (3\text{-}12)$$

where $\Gamma(r,\vartheta,z)$ is the scalar functional of v_r, v_ϑ and v_z defined by the tensor operation

$$\Gamma = \{\, 2 \text{ trace } (\underline{\underline{D}} o \underline{\underline{D}}) \,\}^{1/2} \qquad (3\text{-}13)$$

These considerations are still very general. Let us examine an important and practical simplification. The Ostwald-de Waele model for two-parameter "power law fluids" assumes that the apparent viscosity satisfies

$$N(\Gamma) = k \, \Gamma^{n-1} \qquad (3\text{-}14)$$

where the "consistency factor" k and the "fluid exponent" n are constants. Power law fluids are "pseudoplastic" when $0 < n < 1$, Newtonian when $n = 1$, and "dilatant" when $n > 1$. Most drilling fluids are pseudoplastic. In the limit ($n=1$, $k=\mu$), Equation 3-14 reduces to the Newtonian model with $N(\Gamma) = \mu$, where μ is the laminar viscosity; in this case, stress is directly proportional to shear rate.

On the other hand, when n and k take on general values, the apparent viscosity function becomes somewhat complicated. For isotropic, rotating flows without velocity dependence on the azimuthal coordinate ϑ, the function Γ in Equation 3-14 takes the form

$$\Gamma = [\, (\partial v_z/\partial r)^2 + r^2 \, (\partial\{v_\vartheta/r\}/\partial r)^2 \,]^{1/2} \qquad (3\text{-}15)$$

as we will show, so that Equation 3-14 becomes

$$N(\Gamma) = k \, [(\partial v_z/\partial r)^2 + r^2 \, (\partial\{v_\vartheta/r\}/\partial r)^2 \,]^{(n-1)/2} \qquad (3\text{-}16)$$

This expression for the apparent viscosity reduces to the conventional $N(\Gamma) = k \, (\partial v_z/\partial r)^{(n-1)}$ for "axial only" flows without rotation, and to the formula $N(\Gamma) = k \, (r\partial\{v_\vartheta/r\}/\partial r)^{(n-1)}$ for "rotation only" viscometer flows without axial velocity.

When both axial and circumferential velocities are present, as in annular flows with drillstring rotation, neither of these simplifications apply. This leads to mathematical difficulty. Even though "$v_\vartheta(max)$" is known from the rotation rate, the magnitude of the nondimensional "$v_\vartheta(max)/v_z(max)$" ratio cannot be accurately estimated because v_z is highly sensitive to n, k and pressure drop. Thus, it is impossible to determine beforehand whether or not rotation effects will be weak; simple "axial flow only" formulas cannot be used a priori.

The Newtonian flow - an exact solution to the Navier Stokes equations - is considered first. Then an approximate solution for pseudoplastic and dilatant power law fluids is developed for more general n's. We will derive closed form results for rotating flows using Equation 3-16 in its entirety. We assume a narrow annulus, however, but no further simplifications are taken. Because the mathematical manipulations are complicated, the Newtonian limit is examined first to gain insight into the general case. This is instructive because the physical differencess between Newtonian and power law fluids will also be highlighted.

The annular geometry is shown in Figure 2-1. A drillpipe (or casing) and borehole combination is inclined at an angle α relative to the ground, with $\alpha=0^\circ$ for horizontal wells and $\alpha=90^\circ$ for vertical ones. "z" denotes any point within the drillpipe or annular fluid; Section "AA" is cut normal to the local z axis. Figure 2-2 again resolves the vertical body force at "z" due to gravity into components parallel and perpendicular to the axis. Figure 3-1 further breaks the latter into vectors in the radial and azimuthal directions of the cylindrical coordinate system at Section "AA".

Physical assumptions about the drillstring and borehole flow relative to these coordinates are developed next. Again the problem is solved *exactly* for Newtonian flows for all annular diameters. Then the solution for power law fluids - in the narrow annulus limit - is addressed. The physical and mathematical consistency of these solutions will be evaluated, and applications formulas and detailed calculations are given.

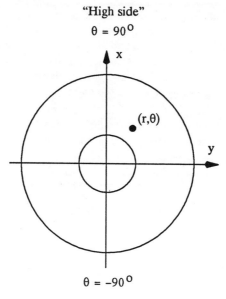

"High side"
$\theta = 90^{O}$

(r,θ)

$\theta = -90^{O}$

"Low side"

Concentric circular pipe and hole

Expanded view

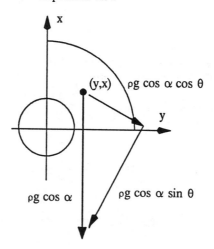

(y,x) $\rho g \cos \alpha \cos \theta$

$\rho g \cos \alpha$ $\rho g \cos \alpha \sin \theta$

Free body diagram in
(r,θ,z) coordinates

Figure 3-1. Gravity Vector Components.

EXACT SOLUTION FOR NEWTONIAN FLOWS

For Newtonian flows, the stress is linearly proportional to the shear rate; the proportionality constant is the laminar fluid viscosity μ. We assume for simplicity that μ is constant. In high shear gradient flows with non-negligible heat generation, μ would depend on temperature, which in turn affects the velocity; in this event, an additional coupled energy equation would be needed. For the problem at hand, Equations 3-1 to 3-3 become

Momentum equation in r:

$$\rho\ (\partial v_r/\partial t + v_r\ \partial v_r/\partial r + v_\vartheta/r\ \partial v_r/\partial \vartheta - v_\vartheta^2/r + v_z\ \partial v_r/\partial z)\ =$$
$$= F_r - \partial p/\partial r + \mu\ \{\partial^2 v_r/\partial r^2 + 1/r\ \partial v_r/\partial r - v_r/r^2$$
$$+ 1/r^2\ \partial^2 v_r/\partial \vartheta^2 - 2/r^2\ \partial v_\vartheta/\partial \vartheta + \partial^2 v_r/\partial z^2\} \qquad (3\text{-}17)$$

Momentum equation in ϑ:

$$\rho\ (\partial v_\vartheta/\partial t + v_r\ \partial v_\vartheta/\partial r + v_\vartheta/r\ \partial v_\vartheta/\partial \vartheta + v_r v_\vartheta/r + v_z\ \partial v_\vartheta/\partial z) =$$
$$= F_\vartheta - 1/r\ \partial p/\partial \vartheta + \mu\ \{\partial^2 v_\vartheta/\partial r^2 + 1/r\ \partial v_\vartheta/\partial r - v_\vartheta/r^2$$
$$+ 1/r^2\ \partial^2 v_\vartheta/\partial \vartheta^2 + 2/r^2\ \partial v_r/\partial \vartheta + \partial^2 v_\vartheta/\partial z^2\} \qquad (3\text{-}18)$$

Momentum equation in z:

$$r\ (\partial v_z/\partial t + v_r\ \partial v_z/\partial r + v_\vartheta/r\ \partial v_z/\partial \vartheta + v_z\ \partial v_z/\partial z) =$$
$$= F_z - \partial p/\partial z + \mu\ (\partial^2 v_z/\partial r^2 + 1/r\ \partial v_z/\partial r$$
$$+ 1/r^2\ \partial^2 v_z/\partial \vartheta^2 + \partial^2 v_z/\partial z^2) \qquad (3\text{-}19)$$

In this section, it will also be convenient to rewrite Equation 3-4 in the expanded form

Continuity equation:

$$\partial v_r/\partial r + v_r/r + 1/r\ \partial v_\vartheta/\partial \vartheta + \partial v_z/\partial z = 0 \qquad (3\text{-}20)$$

Now consider the free body diagrams in Figures 2-1, 2-2 and 3-1. Figure 2-1 shows a straight borehole containing a centered, rotating drillstring inclined at an angle α relative to the ground. Figure 2-2, referring to this geometry, resolves the gravity vector **g** into components parallel and perpendicular to the borehole axis. Figure 3-1 applies to the circular cross-section "AA" of Figure 2-1 and introduces local cylindrical coordinates (r,ϑ). The "low side $\vartheta=-90^o$" marks the position where cuttings beds would normally form. The force $\rho g \cos \alpha$ of Figure 2-2 is also resolved into the orthogonal components $\rho g \cos \alpha \sin \vartheta$ and $\rho g \cos \alpha \cos \vartheta$.

Physical assumptions about the flow are now given. First, it is expected that at any section AA along the borehole axis z, the velocity fields will appear to be the same; they are invariant, so z derivatives of v_r, v_ϑ and v_z vanish. Also, since the drillpipe and borehole walls are assumed to be impermeable, $v_r = 0$ throughout (in formation invasion studies, this would not apply). While we do have physically unsteady pipe rotation, note that the use of circular cylindrical coordinates (with constant v_ϑ at the drillstring) renders the mathematical formulation steady. Thus all time derivatives vanish. These assumptions imply

$$-\rho v_\vartheta{}^2/r = F_r - \partial p/\partial r + \mu\{ -2/r^2 \, \partial v_\vartheta/\partial \vartheta \ \} \tag{3-21}$$

$$\rho v_\vartheta/r \, \partial v_\vartheta/\partial \vartheta = F_\vartheta - 1/r \, \partial p/\partial \vartheta + \mu \, \{\partial^2 v_\vartheta/\partial r^2 + 1/r \, \partial v_\vartheta/\partial r \\ - v_\vartheta/r^2 + 1/r^2 \, \partial^2 v_\vartheta/\partial \vartheta^2\} \tag{3-22}$$

$$\rho v_\vartheta/r \, \partial v_z/\partial \vartheta = F_z - \partial p/\partial z + \mu \, (\partial^2 v_z/\partial r^2 + 1/r \, \partial v_z/\partial r \\ + 1/r^2 \, \partial^2 v_z/\partial \vartheta^2) \tag{3-23}$$

$$\partial v_\vartheta/\partial \vartheta = 0 \tag{3-24}$$

Equation 3-24 is instrumental in simplifying Equations 3-21 to 3-23 further. We straightforwardly obtain

$$- \rho v_\vartheta{}^2/r = - \rho g \cos \alpha \sin \vartheta - \partial p/\partial r \tag{3-25}$$

$$0 = - \rho g \cos \alpha \cos \vartheta - 1/r \, \partial p/\partial \vartheta \\ + \mu \, \{\partial^2 v_\vartheta/\partial r^2 + 1/r \, \partial v_\vartheta/\partial r - v_\vartheta/r^2\} \tag{3-26}$$

$$\rho \ v_\vartheta/r \, \partial v_z/\partial \vartheta = \rho g \sin \alpha - \partial p/\partial z \\ + \mu \, (\partial^2 v_z/\partial r^2 + 1/r \, \partial v_z/\partial r + 1/r^2 \, \partial^2 v_z/\partial \vartheta^2) \tag{3-27}$$

where we have additionally substituted the body force components of Figures 2-2 and 3-1. Now, since Equation 3-27 does not explicitly contain ϑ, it follows that v_z is independent of ϑ. Since we had already shown that there is no z dependence, $v_z = v_z(r)$ is a function of r only. Equation 3-27 therefore becomes

$$0 = \rho g \sin \alpha - \partial p/\partial z + \mu \, (\partial^2 v_z/\partial r^2 + 1/r \, \partial v_z/\partial r) \tag{3-28}$$

To achieve further simplicity, we resolve (without loss of generality) the pressure $p(r,\vartheta,z)$ into its component dynamic pressures $P(z)$ and $P^*(r)$, and its hydrostatic contribution, through the separation of variables

$$p(r,\vartheta,z) = P(z) + P^*(r) + z\rho g \sin \alpha - r\rho g \cos \alpha \sin \vartheta \qquad (3\text{-}29)$$

This reduces the governing Navier-Stokes equations to the simpler but mathematically equivalent system

$$\partial^2 v_z/\partial r^2 + 1/r \; \partial v_z/\partial r = 1/\mu \; dP(z)/dz = \text{constant} \qquad (3\text{-}30)$$

$$\partial^2 v_\vartheta/\partial r^2 + 1/r \; \partial v_\vartheta/\partial r - v_\vartheta/r^2 = 0 \qquad (3\text{-}31)$$

$$\rho v_\vartheta^2/r = \partial P^*(r)/\partial r = dP^*(r)/dr \qquad (3\text{-}32)$$

The separation of variables introduced in Equation 3-29 and the explicit elimination of "g" in Equations 3-30 to 3-32 does not mean that gravity is unimportant; the effects of gravity are simply tracked in the $dP(z)/dz$ term of Equation 3-30. The function $P^*(r)$ will depend on the velocity solution to be obtained.

Equations 3-30 to 3-32 are also significant in another respect. The velocity fields $v_z(r)$ and $v_\vartheta(r)$ can be obtained *independently* of each other, despite the nonlinearity of the Navier-Stokes equations, because Equations 3-30 and 3-31 physically uncouple. This decoupling occurs because the nonlinear convective terms in the original momentum equations identically vanish. Equation 3-32 is only applied (after the fact) to calculate the radial pressure field $P^*(r)$ for use in Equation 3-29. This decoupling applies only to Newtonian flows. For non-Newtonian flows, $v_z(r)$ and $v_\vartheta(r)$ are strongly coupled mathematically, and different solution strategies are needed.

This degeneracy with Newtonian flows means that their physical properties will be completely different from those of power law fluids. For Newtonian flows, changes in rotation rate will not affect properties in the axial direction, in contrast to non-Newtonian flows. Cuttings transport recommendations deduced, for example, using water as the working medium, cannot be extrapolated to general drilling fluids having fractional values of n, using any form of dimensional analysis. Similarly, observations for power law fluids need not apply to water.

This uncoupling was, apparently, first observed by Savins and Wallick (1966). This author is indebted to J. Savins for bringing this earlier result to his attention. Savins and Wallick noted that in Newtonian flows, no coupling among the discharge rate, axial pressure gradient, and relative motion and torque through the viscosity coefficient exists. But we emphasize that the coupling between v_z and v_ϑ reappears in eccentric geometries even for Newtonian flows.

Because Equations 3-30 and 3-31 are linear, it is possible to solve for the complete flowfield using exact classical methods. We will give all required solutions without proof, since they can be verified by direct substitution. For the *inside* of the drillpipe, the axial flow solution to Equation 3-30 satisfying no-slip conditions at the pipe radius $r = R_P$ and zero shear stress at the centerline $r = 0$ is

$$v_z(r) = (r^2 - R_P^2)/4\mu \ dP(z)/dz \tag{3-33}$$

Similarly, the rotating flow solution to Equation 3-31 satisfying bounded flow at $r = 0$ and $v_\vartheta/r = w$ at $r = R_P$ is

$$v_\vartheta = wr \tag{3-34}$$

This is just the expected equation for solid body rotation. Here, "w" is a constant drillstring rotation rate. These velocity results, again, can be linearly superposed despite the nonlinearity of the underlying equations.

Now, let L denote the length of the pipe, P_{mp} be the constant pressure at the "mudpump" $z = 0$, and P^- be the drillpipe pressure at $z = L$ just upstream of the bit nozzles. Direct integration of Equation 3-32 and substitution in Equation 3-29 yield the complementary solution for pressure

$$p(r,\vartheta,z) = P_{mp} + (P^- - P_{mp}) \ z/L \ + \rho w^2 r^2/2$$
$$+ \rho g(z \sin \alpha - r \cos \ \alpha \sin \ \vartheta) + \text{constant} \tag{3-35}$$

For the annular region between the rotating drillstring and the stationary borehole wall, the solution of Equation 3-30 satisfying no-slip conditions at the pipe radius $r = R_P$ and at the borehole radius $r = R_B$ is

$$v_z(r) = \{r^2 - R_P^2 + (R_B^2 - R_P^2) \ (\log r/R_P)/\log R_P/R_B\} \ 1/4\mu \ dP(z)/dz \tag{3-36}$$

where "log" denotes the natural logarithm. The solution of Equation 3-31 satisfying $v_\vartheta = 0$ at $r = R_B$ and $v_\vartheta = wr$ at $r = R_P$ is

$$v_\vartheta(r) = wR_P(R_B/r - r/R_B)/(R_B/R_P - R_P/R_B) \tag{3-37}$$

Now let P^+ be the pressure at $z = L$ just outside of the bit nozzles, and P_{ex} be the surface exit pressure at $z = 0$. The corresponding solution for pressure from Equation 3-32 is

$$p(r,\vartheta,z) = P^+ + (P_{ex} - P^+) (L-z)/L + \rho g(z \sin \alpha - r \cos \alpha \sin \vartheta)$$
$$+ \rho w^2 R_P^2 \{-0.5(R_B/r)^2 + 0.5(r/R_B)^2 - 2 \log (r/R_B) + constant\}/$$
$$(R_B/R_P - R_P/R_B)^2 \tag{3-38}$$

Observe that the pressure $p(r,\vartheta,z)$ depends on all three coordinates, even though $v_z(r)$ depends only on r. The pressure gradient $\partial p/\partial r$, for example, throws cuttings through centrifugal force; it likewise depends on r,ϑ and z, and also on ρ, g and α. It may be an important correlation parameter in cuttings transport and bed formation studies.

The additive constants in Equations 3-35 and 3-38 have no dynamical significance. Equations 3-33 to 3-38 describe completely and *exactly* the internal drillpipe flow and the external annular borehole flow. No geometrical simplifications have been made. The solution applies to an inclined, centered drillstring rotating at a constant angular rate w, but it is restricted to a Newtonian fluid.

Again, these concentric solutions show that in the Newtonian limit, the velocities $v_z(r)$ and $v_\vartheta(r)$ uncouple; this is not the case for eccentric flows. And this is not so with non-Newtonian drilling fluids in either concentric or eccentric geometries. The analysis methods developed here will now be extended to account for the coupling inherent in power law flows.

NARROW ANNULUS POWER LAW SOLUTION

For general non-Newtonian flows, the Navier-Stokes equations (see Equations 3-17 to 3-19) do not apply; direct recourse to Equations 3-1 to 3-3 must be made. However, many of the physical assumptions used and justified above still hold. If we again assume a constant density flow, and also that velocities do not vary with z, ϑ and t, and also that $v_r=0$, we again obtain Equation 3-24. This implies mass conservation. It leads to further simplifications in Equations 3-1 to 3-3, and in the tensor definitions given by Equations 3-5 to 3-14. The result is the reduced system of momentum equations

$$0 = \rho g \sin \alpha - \partial p/\partial z + 1/r\ \partial(Nr\ \partial v_z/\partial r)/\partial r \tag{3-39}$$

$$0 = -\rho g \cos\alpha \cos\vartheta - 1/r\ \partial p/\partial\vartheta + 1/r^2\ \partial(Nr^3\ \partial(v_\vartheta/r)/\partial r)/\partial r \tag{3-40}$$

$$- \rho\ v_\vartheta^2/r = - \rho g \cos \alpha \sin\vartheta - \partial p/\partial r$$
$$+ 1/r\ \partial(Nr\ \partial(v_\vartheta/r)/\partial r)/\partial\vartheta + \partial(N\ \partial v_z/\partial r)/\partial z \tag{3-41}$$

At this point, we introduce the same separation of variables for pressure used previously for Newtonian flows, that is Equation 3-29, so that Equations 3-39 to 3-41 become

$$0 = -\partial P/\partial z + 1/r \; \partial(Nr \; \partial v_z/\partial r)/\partial r \qquad (3\text{-}42)$$

$$0 = \partial(Nr^3 \; \partial(v_\vartheta/r)/\partial r)/\partial r \qquad (3\text{-}43)$$

$$-\rho v_\vartheta^2/r = -\partial P^*/\partial r + 1/r \; \partial(Nr \; \partial(v_\vartheta/r)/\partial r)/\partial\vartheta + \partial(N \; \partial v_z/\partial r)/\partial z \qquad (3\text{-}44)$$

Of course, the $P^*(r)$ applicable to non-Newtonian flows will follow from the solution to Equation 3-44; Equations 3-35 and 3-38 for Newtonian flows do not apply.

Since ϑ does not explicitly appear in Equation 3-44, v_z and v_ϑ do not depend on ϑ; they do not depend on z either, as previously assumed. Thus, all partial derivatives with respect to ϑ and z vanish. Without approximation, the final set of ordinary differential equations (ODE) takes the form

$$1/r \; d(Nr \; dv_z/dr)/dr = dP/dz = \text{constant} \qquad (3\text{-}45)$$

$$d(Nr^3 \; d(v_\vartheta/r)dr)/dr = 0 \qquad (3\text{-}46)$$

$$dP^*/dr = \rho \; v_\vartheta^2/r \qquad (3\text{-}47)$$

where $N(\Gamma)$ is the complete velocity functional given in Equation 3-16. The application of Equation 3-16 couples our axial and azimuthal velocities, and is the source of mathematical complication.

The solution to Equations 3-45 to 3-47 may appear to be simple. For example, the unknowns v_ϑ and v_z are governed by two second-order ODEs, namely, Equations 3-45 and 3-46; the four constants of integration are completely determined by four no-slip conditions at the rotating drillstring surface and the stationary borehole wall. And, the radial pressure (governed by Equation 3-47) is obtained after the fact only, once v_ϑ is available.

In reality, the difficulty lies with the fact that Equations 3-45 and 3-46 are nonlinearly coupled through Equation 3-16. It is not possible to solve for either v_z and v_ϑ sequentially, as we did for the "simpler" Navier-Stokes equations. Because the actual physical coupling is strong at the leading order, it is incorrect to solve for non-Newtonian effects using perturbation methods, say, expanded about the degenerate and decoupled Newtonian solutions. The method described here required tedious trial and error; twenty-four ways to implement no-slip conditions were possible, and not all yielded integrable equations.

We successfully derived closed form, explicit, analytical solutions for the coupled velocity fields. However, the desire for closed solutions required an additional "narrow annulus" assumption. Still, the resulting solutions are useful since they yield explicit answers for rotating flows, thus providing key physical insight into the role of different flow parameters. The method devised for arbitrary n does *not* apply to the Newtonian limit n = 1, for which solutions are already available. But in the n --> 1± limit of our power law results, we will show that we importantly recover our Navier-Stokes solution. Thus, the physical dependence on n is continuous, and the results obtained in this chapter cover all values of n. With these preliminary remarks said and done, we proceed with the analysis.

Let us multiply Equation 3-45 by r throughout. Next, integrate the result; also integrate Equation 3-46 once with respect to r to yield

$$Nr \, dv_z/dr = r^2/2 \, dP/dz + E_1 \tag{3-48}$$

$$Nr^3 \, d(v_\vartheta/r)dr = E_2 \tag{3-49}$$

where E_1 and E_2 are integration constants. Division of Equation 3-48 by Equation 3-49 gives a result (independent of the apparent viscosity $N(\Gamma)$) relating v_z to v_ϑ/r, namely,

$$dv_z/dr = (r^4/2 \, dP/dz + E_1 r^2)/E_2 \, d(v_\vartheta/r)dr \tag{3-50}$$

At this point, it will be more convenient to introduce the angular velocity variable

$$\Omega(r) = v_\vartheta/r \tag{3-51}$$

for subsequent use. Substitution of the isotropic tensor elements $\underline{\underline{D}}$ in Equation 3-13 leads to

$$\begin{aligned} \Gamma \quad &= \{ \, 2 \, \text{trace} \, (\underline{\underline{D}} \circ \underline{\underline{D}}) \, \}^{1/2} \\ &= [\, (\partial v_z/\partial r)^2 + r^2 \, (\partial\{v_\vartheta/r\}/\partial r)^2 \,]^{1/2} \end{aligned} \tag{3-52}$$

so that the power law apparent viscosity given by Equation 3-14 becomes

$$N(\Gamma) = k \, [(\partial v_z/\partial r)^2 + r^2 \, (\partial\{v_\vartheta/r\}/\partial r)^2 \,]^{(n-1)/2} \tag{3-53}$$

These results were stated without proof in Equations 3-15 and 3-16. Now combine Equations 3-49 and 3-53 so that

$$k \, [(\partial v_z/\partial r)^2 + r^2 \, (\partial\Omega/\partial r)^2 \,]^{(n-1)/2} \, d\Omega/dr = E_2/r^3 \tag{3-54}$$

If dv_z/dr is eliminated using Equation 3-50, we obtain after lengthy manipulations

$$d\Omega/dr = (E_2/k)^{1/n} \, [r^{(2n+4)/(n-1)}$$
$$+ r^{(4n+2)/(n-1)} \, \{(E_1 + r^2/2 \, dP/dz)/E_2\}^2 \,]^{(1-n)/2n} \tag{3-55}$$

Next, integrate Equation 3-55 over the interval (r, R_B) where R_B is the borehole radius. If we apply the first no-slip boundary condition

$$\Omega(R_B) = 0 \tag{3-56}$$

(there are four no-slip conditions altogether) and invoke the Mean Value Theorem of differential calculus, using the arithmetic average as an appropriate mean, we obtain

$$\Omega(r) = (E_2/k)^{1/n} \, (r-R_B) \, [((r+R_B)/2) \,^{(2n+4)/(n-1)} \tag{3-57}$$
$$+((r+R_B)/2)^{(4n+2)/(n-1)} \, \{(E_1 + (r+R_B)^2/8 \, dP/dz)/E_2\}^2 \,]^{(1-n)/2n}$$

which still contains the unknown constants E_1 and E_2. At this point, though, it is *not* useful to apply any of the remaining three no-slip conditions.

So, we turn our attention to v_z. We can derive a differential equation independent of Ω by combining Equations 3-50, 3-51 and 3-55 as follows,

$$dv_z/dr \quad = (r^4/2 \, dP/dz + E_1 r^2)/E_2 \, d\Omega/dr$$
$$= r^2(E_1 + r^2/2 \, dP/dz)/E_2 \, \mathbf{X} \, (E_2/k)^{1/n} \, [r^{(2n+4)/(n-1)}$$
$$+ r^{(4n+2)/(n-1)} \, \{(E_1 + r^2/2 \, dP/dz)/E_2\}^2 \,]^{(1-n)/2n} \tag{3-58}$$

We next integrate Equation 3-58 over (R_P, r), where R_P is the drillpipe radius, subject to the second no-slip condition

$$v_z(R_P) = 0 \tag{3-59}$$

An integration similar to that used for Equation 3-55, again invoking the Mean Value Theorem, leads to a result analogous to Equation 3-57. That is, we obtain

$$v_z(r) = \quad ((r+R_P)/2)^2(E_1 + ((r+R_P)/2)^2/2 \, dP/dz)/E_2 \, \mathbf{X}$$
$$(E_2/k)^{1/n} \, [((r+R_P)/2)^{(2n+4)/(n-1)} +((r+R_P)/2)^{(4n+2)/(n-1)}$$
$$\{(E_1 + ((r+R_P)/2)^2/2 \, dP/dz)/E_2\}^2 \,]^{(1-n)/2n} \, (r - R_P) \tag{3-60}$$

Very useful results are obtained if we now apply the third no-slip condition

$$v_z(R_B) = 0 \tag{3-61}$$

With this constraint, Equation 3-60 leads to an unwieldy combination of terms,

$$
\begin{aligned}
0 = &((R_B+R_P)/2)^2(E_1 + ((R_B+R_P)/2)^2/2 \; dP/dz)/E_2 \; \times \\
&(E_2/k)^{1/n} \; [((R_B+R_P)/2)^{(2n+4)/(n-1)} + ((R_B+R_P)/2)^{(4n+2)/(n-1)} \\
&\{E_1 + ((R_B+R_P)/2)^2/2 \; dP/dz)/E_2\}^2 \;]^{(1-n)/2n} \; (R_B - R_P)
\end{aligned}
\tag{3-62}
$$

At first, this appears hopelessly complicated. But if we observe that the quantity contained within the square brackets "[ ]" is positive definite, and that $(R_B - R_P)$ is nonzero, it follows that the left hand side "0" can be obtained only if

$$E_1 = -(R_B+R_P)^2/8 \; dP/dz \tag{3-63}$$

holds identically. The remaining integration constant E_2 is determined from the last of the four no-slip conditions

$$\Omega(R_P) = w \tag{3-64}$$

Equation 3-64 requires fluid at the pipe surface to move with the rotating surface. Here, without loss of generality, we follow the sign convention $w < 0$ for the constant drillstring angular rotation speed assumed given. Combination of Equations 3-57, 3-63 and 3-64, after lengthy manipulations, leads to the surprisingly simple result that

$$E_2 = k \; (w/(R_P - R_B))^n \; ((R_P+R_B)/2)^{n+2} \tag{3-65}$$

With all four no-slip conditions applied, the four integration constants and hence the analytical solution for our power law formulation, are completely determined. We next perform some validation checks before deriving applications formulas.

ANALYTICAL VALIDATION

Different analytical procedures were required for Newtonian and power law flows. This arises from the decoupling between axial and circumferential velocities in the singular n=1 limit. On physical grounds, we expect that the power law solution, if correct, would behave "continuously" through n = 1 as the fluid passes from dilatant to pseudoplastic states. That is, the solution should change smoothly when n varies from 1-δ to 1+δ where $|\delta| \ll 1$ is a small number. This continuous dependence and physical consistency will be demonstrated next. This validation also guards against error, given the quantity of algebraic manipulations involved.

The formulas derived above for general power law fluids will be checked against exact Newtonian results where k=μ and n=1. For consistency, we will take the narrow annulus limit of those formulas, a geometric approximation used in the power law derivation. We will demonstrate that the closed form results obtained for non-Newtonian fluids are indeed "continuous in n" through the singular point n=1.

We first check our results for the stresses $S_{r\vartheta}$ and $S_{\vartheta r}$. From Equations 3-5, 3-9 and 3-57, we find that

$$S_{r\vartheta} = S_{\vartheta r} = k \; (w/(R_P - R_B))^n \; ((R_P+R_B)/2)^{n+2} \; r^{-2} \tag{3-66}$$

In the limit k=μ and n=1, Equation 3-66 for power law fluids reduces to

$$S_{r\vartheta} = S_{\vartheta r} = \mu \; w/\{(R_P - R_B)r^2\} \; \text{X} \; ((R_P+R_B)/2)^3 \tag{3-67}$$

On the other hand, the definition $S_{r\vartheta} = S_{\vartheta r} = \mu \; d\Omega/dr$ inferred from Equations 3-5 and 3-9 becomes, using Equations 3-37 and 3-51 for Newtonian flow,

$$S_{r\vartheta} = S_{\vartheta r} = \mu \; w/\{(R_P - R_B)r^2\} \; \text{X} \; 2(R_P R_B)^2/(R_P + R_B) \tag{3-68}$$

Are the second factors " $((R_P+R_B)/2)^3$ " and "$2(R_P R_B)^2/(R_P + R_B)$ " in Equations 3-67 and 3-68 consistent? If we evaluate these expressions in the narrow annulus limit, setting $R_P = R_B = R$, we obtain R^3 in *both* cases, providing the required validation. This consistency holds for all values of dP/dz.

For our second check, consider the power law stresses S_{rz} and S_{zr} obtained from Equations 3-5, 3-10 and 3-58, that is,

$$
\begin{aligned}
S_{rz} = S_{zr} \;\; &= E_1/r + 0.5 \; r \; dP/dz \\
&= \{0.5 \; r - (R_P+R_B)^2/(8r)\} \; dP/dz
\end{aligned}
\tag{3-69}
$$

The corresponding formula in the Newtonian limit is

$$S_{rz} = S_{zr} \quad = \mu \, dv_z/dr$$
$$= \{0.5r - (R_P^2 - R_B^2)/(4r \log R_P/R_B)\} \, dP/dz \qquad (3\text{-}70)$$

where we have used Equation 3-36. Is "$(R_P+R_B)^2/8$" consistent with "$(R_P^2 - R_B^2)/(4 \log R_P/R_B)$"? As before, consider the narrow annulus limit, setting $R_P = R_B = R$. The first expression easily reduces to $R^2/2$. For the second, we expand $\log R_B/R_P = \log \{1 + (R_B-R_P)/R_P\} = (R_B-R_P)/R_P$ and retain only the first term of the Taylor expansion. Direct substitution yields $R^2/2$ again. Therefore, Equations 3-69 and 3-70 are consistent for all rotation rates w. Thus, from our checks on both S_{rz} and $S_{r\vartheta}$, we find good physical consistency and consequently reliable algebraic computations.

DIFFERENCES BETWEEN NEWTONIAN
AND POWER LAW FLOWS

Equations 3-29, 3-51, 3-57, 3-60, 3-63 and 3-65 completely specify the velocity fields v_z and $v_\vartheta = r\Omega(r)$ as functions of wellbore geometry, fluid rheology, pipe inclination, rotation rate, pressure gradient and gravity. We emphasize that Equation 3-47, which is to be evaluated using the non-Newtonian solution for v_ϑ, provides only a partial solution for the complete radial pressure gradient. The remaining part is obtained by adding the "$-\rho g \cos \alpha \sin \vartheta$" contribution of Equation 3-29. As in Newtonian flows, the pressure and its spatial gradients depend on all the coordinates r, ϑ and z, and the parameters ρ, g and α.

There are fundamental differences between these solutions and the Newtonian ones. For example, in the latter, the solutions for v_z and v_ϑ completely decouple despite the nonlinearity of the Navier-Stokes equations. The governing equations in fact become linear. But for power law flows, both v_z and v_ϑ remain highly coupled and nonlinear. In this sense, Newtonian results are singular; but the degeneracy would disappear for eccentric geometries when the convective terms reappear. Cuttings transport experimenters working with *concentric Newtonian* flows will *not* be able to extrapolate their findings to more practical geometries or fluids. This subject will be discussed in greater detail in Chapter 5.

Also, the expression for v_z in the Newtonian limit is directly proportional to dP/dz; but as Equation 3-60 for power law fluids shows, the dependence of v_z on pressure gradient is a nonlinear one. Similarly, while Equation 3-37 shows that v_ϑ is directly proportional to the rotation rate w, Equations 3-51, 3-57 and 3-65 illustrate a more complicated nonlinear dependence for power law fluids. It is important to emphasize that, for a fixed annular flow geometry and a Newtonian flow, v_z depends only on dP/dz and not w, and v_ϑ depends only on w and not dP/dz. But for power law flows, v_z and v_ϑ each depend on both dP/dz and w. Thus, "axial quantities" like net annular volume flow rate cannot be calculated without considering both dP/dz and w.

Interestingly, though, the stress formulas for $S_{r\vartheta}$ and S_{rz} in the non-Newtonian case preserve their "independence" as in Newtonian flows. That is, $S_{r\vartheta}$ depends only on w and not dP/dz, while S_{rz} depends only on dP/dz and not w (see Equations 3-71 to 3-74 below). The power law stress values themselves, of course, are different from the Newtonian counterparts. And, the "maximum stress" $(S_{r\vartheta}^2 + S_{rz}^2)^{1/2}$, important in borehole stability and cuttings bed erosion, depends on both w and dP/dz as it does in Newtonian flow.

An important question is the significance of rotation in practical calculations. Can "w" be safely neglected in drilling and cementing applications? This depends on a nondimensional ratio of circumferential to axial momentum flux. While the "maximum v_ϑ" is easily obtained as "$w_{rpm} \times R_P$", the same estimate for v_z is difficult to produce since axial velocity is sensitive to both n and k, not to mention v_ϑ and dP/dz. In general, one needs to consider the full problem without approximation.

Of course, since the analytical solution is available anyway, the use of approximate "axial flow only" solutions is really a moot point. The power law results and the formulas to be derived next are "explicit" in that they require no iteration. And although the software described is written in Fortran, our equations are just as easily programmed on calculators. The important dependence of annular flows on "w" will be demonstrated in calculated results.

MORE APPLICATIONS FORMULAS

The geometry of the present problem renders all stress tensor components except $S_{r\vartheta}$, $S_{\vartheta r}$, S_{zr} and S_{rz} zero. From our power law results, the required formulas for viscous stress can be shown to be

$$S_{r\vartheta} = S_{\vartheta r} = k \ (w/(R_P - R_B))^n \ ((R_P + R_B)/2)^{n+2} \ r^{-2} \tag{3-71}$$

$$S_{rz} = S_{zr} = E_1/r + 0.5 \ r \ dP/dz$$
$$= \{0.5 \ r \ -(R_P + R_B)^2/(8r)\} \ dP/dz \tag{3-72}$$

Their Newtonian counterparts take the form

$$S_{r\vartheta} = S_{\vartheta r} = \mu \, w/\{(R_P - R_B)r^2\} \text{ X } 2(R_P R_B)^2/(R_P + R_B) \tag{3-73}$$

$$\begin{aligned} S_{rz} = S_{zr} &= \mu \, dv_z/dr \\ &= \{0.5r - (R_P^2 - R_B^2)/(4r \log R_P/R_B)\} \, dP/dz \end{aligned} \tag{3-74}$$

In studies on borehole erosion, annular velocity plays an important role, since drilling mud carries abrasive cuttings. The magnitude of fluid shear stress may also be important in unconsolidated sands where tangential surface forces assist in wall erosion. Stress considerations also arise in cuttings bed transport analysis in highly deviated or horizontal holes (see Chapter 5). The individual components can be obtained by evaluating Equations 3-71 and 3-72 at $r = R_B$ for power law fluids, and Equations 3-73 and 3-74 for Newtonian fluids. And since these stresses act in orthogonal directions, the "maximum stress" can be obtained by writing

$$S_{max}(R_B) = \{S_{r\vartheta}^2(R_B) + S_{rz}^2(R_B)\}^{1/2} \tag{3-75}$$

The shear force associated with this stress acts in a direction offset from the borehole axis by an angle

$$\vartheta \text{ max shear } = \text{arc tan } \{S_{r\vartheta}(R_B)/S_{rz}(R_B)\} \tag{3-76}$$

Opposing the erosive effects of shear may be the stabilizing effects of hydrostatic and dynamic pressure. Explicit formulas for the pressures $P(z)$, $P^*(r)$ and the hydrostatic background level were given earlier.

To obtain the corresponding elements of the deformation tensor, we rewrite Equation 3-5 in the form

$$\underline{\underline{D}} = \underline{\underline{S}} / 2N(\Gamma) \tag{3-77}$$

and substitute S_{rz} or $S_{r\vartheta}$ as required. In the Newtonian case, $N(\Gamma) = \mu$ is the laminar viscosity; for power law fluids, Equation 3-16 applies. Stresses are important to transport problems; fluid deformations are useful for the kinematic studies often of interest to rheologists.

The annular volume flow rate Q depends nonlinearly on pressure gradient; the exact dependence is important in determining mud pump power requirements and cuttings transport capabilities of the drilling fluid. It is obtained by evaluating

$$Q = \int_{R_P}^{R_B} v_z(r)\ 2\pi r\ dr \qquad\qquad (3\text{-}78)$$

In the above integrand, Equation 3-36 for $v_z(r)$ must be used for Newtonian flows, while Equation 3-60 would apply to power law fluids.

Borehole temperature may play an important role in drilling. Problem areas include formation temperature interpretation and mud thermal stability (e.g., the "thinning" of oil base muds with temperature limits cuttings transport efficiency). Many studies do not consider the effects of heat generation by internal friction, which may be non-negligible; in closed systems, temperature increases over time may be significant. Ideally, temperature effects due to fluid type and cumulative effects related to total circulation time should be identified.

The starting point is the equation describing energy balances within the fluid; that is, the PDE for the temperature field $T(r,\vartheta,z,t)$. Even if the velocity field is steady, temperature effects will typically not be, since irreversible thermodynamic effects cause continual increases of T with time. If temperature increases are large enough, the changes of viscosity, consistency factor and fluid exponent as functions of T must be considered. Then the momentum and energy equations will be coupled. We will not consider this complicated situation, so that the velocity fields can be obtained independently of T. For Newtonian flows, we have $n=1$ and $k=\mu$. The temperature field satisfies

$$
\begin{aligned}
\rho c\ (\partial T/\partial t + v_r\ \partial T/\partial r &+ v_\vartheta/r\ \partial T/\partial\vartheta + v_z\ \partial T/\partial z) = \\
&= K\ [\ 1/r\ \partial(r\ \partial T/\partial r)/\partial r + 1/r^2\ \partial^2 T/\partial\vartheta^2 + \partial^2 T/\partial z^2\ ] \\
&\quad + 2\mu\{(\partial v_r/\partial r)^2 + [1/r\ (\partial v_\vartheta/\partial\vartheta\ + v_r)]^2 + (\partial v_z/\partial z)^2\} \\
&\quad + \mu\ \{(\partial v_\vartheta/\partial z + 1/r\ \partial v_z/\partial\vartheta)^2 + (\partial v_z/\partial r\ + \partial v_r/\partial z)^2 \\
&\quad + [1/r\partial v_r/\partial\vartheta\ + r\ \partial(v_\vartheta/r)/\partial r]^2\} + rQ^*
\end{aligned}
\qquad (3\text{-}79)
$$

where c is the heat capacity, K is the thermal conductivity and Q^* is an energy transmission function.

The bold "right hand side" terms are positive definite and represent the heat produced by internal fluid friction. These irreversible thermodynamic effects are referred to collectively as the "dissipation function" or "heat generation function". The dissipation function Φ is in effect a distributed heat source within the moving fluid medium. If we employ the same assumptions as used in our solution of the Navier-Stokes equations for Newtonian flows, this expression reduces to

$$\Phi = \mu\ \{(\partial v_z/\partial r)^2 + r^2(\partial\Omega/\partial r)^2\} > 0 \qquad\qquad (3\text{-}80)$$

which can be easily evaluated using Equations 3-36, 3-37 and 3-51. It is important to recognize that Φ depends on spatial velocity gradients only, and not on velocity magnitudes. In a closed system, the fact that $\Phi > 0$ leads to increases of temperature in time if the borehole walls cannot conduct heat away quickly.

Equations 3-79 and 3-80 assume Newtonian flow. For general fluids, it is possible to show that the dissipation function now takes the specific form

$$\Phi = S_{rr} \, \partial v_r/\partial r + S_{\vartheta\vartheta} \, 1/r \, (\partial v_\vartheta/\partial\vartheta + v_r)$$
$$+ S_{zz} \, \partial v_z/\partial z + S_{r\vartheta} \, [r \, \partial(v_\vartheta/r)/\partial r + 1/r \, \partial v_r/\partial\vartheta]$$
$$+ S_{rz} \, (\partial v_z/\partial r + \partial v_r/\partial z) + S_{\vartheta z} \, (1/r \, \partial v_z/\partial\vartheta + \partial v_\vartheta/\partial z) \qquad (3\text{-}81)$$

The geometrical simplifications used earlier simplifies Equation 3-81 to

$$\Phi = k \, \{(\partial v_z/\partial r)^2 + r^2(\partial\Omega/\partial r)^2\}^{(n+1)/2} > 0 \qquad (3\text{-}82)$$

In the Newtonian limit with $k=\mu$ and $n=1$, Equation 3-82 consistently reduces to Equation 3-80. Equations 3-55 and 3-58 are used to evaluate the expression for Φ above. As before, Φ depends upon velocity gradients only and not magnitudes; it largely arises from high shear at solid boundaries.

DETAILED CALCULATED RESULTS

The power law results were encoded in a Fortran algorithm designed to provide a suite of output "utility" solutions for any set of input data. These may be useful in determining operationally important quantities like volume flow rate and axial speed. But they also provide research utilities needed, for example, to correlate experimental cuttings transport data or interpret formation temperature data.

The core code resides in thirty lines of Fortran. It runs on a "stand alone" basis or as an embedded subroutine for specialized applications. The formulas used are also programmable on calculators.

Inputs include pipe or casing outer diameter, borehole diameter, axial pressure gradient, rotation rate, fluid exponent n, and consistency factor k. Outputs include tables and ASCII character plots versus "r" for a number of useful functions. These are

    o Axial velocity $v_z(r)$
    o Circumferential velocity $v_\vartheta(r)$
    o Fluid rotation rate $w(r)$ ("local rpm")
    o Total absolute speed

o Angle between $v_z(r)$ and $v_{\vartheta}(r)$
o Axial velocity gradient $dv_z(r)/dr$
o Azimuthal velocity gradient $dv_{\vartheta}(r)/dr$
o Angular velocity gradient $dw(r)/dr$
o Radial pressure gradient
o Apparent viscosity versus "r"
o Local frictional heat generation
o All stress tensor components
o Maximum wellbore stress
o All deformation tensor components

We emphasize that the "radial pressure gradient" above refers to the partial contribution in Equation 3-47 which depends on "r" only. For the complete gradient, Equation 3-29 shows that the term " $-\rho g \cos\alpha \sin\vartheta$ " must be appended to the value calculated here. This contribution depends on ρ, g, α and ϑ.

In addition to the foregoing arrays, total annular volume flow rate and radial averages of all of the above quantities are computed. Before proceeding to detailed computations, let us compare our concentric, rotating pipe, *narrow annulus* results in the limit of zero rotation with with an exact solution.

Example 1: East Greenbriar No. 2

A mud hydraulics analysis was performed for "East Greenbriar No. 2" using a computer program offered by a service company. This program, which applies to nonrotating flows only, is based on the *exact* Fredrickson and Bird (1958) solution.

In this example, the drillpipe outer radius is 2.5", the borehole radius is 5.0", the axial pressure gradient is 0.00389 psi/ft, the fluid exponent is 0.724, and the consistency factor is 0.268 lbf $\sec^{0.724}$ / (100 ft^2) (that is, 0.1861 x 10^{-4} lbf $\sec^{0.724}$ /in^2 in the units employed by our program). The exact results computed using this data are an annular volume flow rate of 400 gal/min, and an average axial speed of 130.7 ft/min. The same input data was used in our program, with an assumed drillstring "rpm" of 0.001. We computed 373.6 gal/min and 126.9 ft/min for this nonrotating flow, agreeing to within 7% for the not-so-narrow annulus.

Our model was designed, of course, to include the effects of drillstring rotation. We first considered an extremely large rpm of 300, with the same pressure gradient, to evaluate qualitative effects. The corresponding results were 526.1 gal/min and 175.6 ft/min. The ratio of the average circumferential speed to the average axial speed is 1.06, indicating that rotational effects are important.

At 150 rpm, our volume flow rate of 458.7 gal/min exceeds 373.6 gal/min by 23%. In this case, the ratio of average circumferential speed to axial speed is still a non-negligible 65%. These results suggest that static models tend to overestimate the pressure requirements needed by a rotating drillstring to produce a prescribed flow rate. Our hydraulics model indicates that the inclusion of rotational effects, for a fixed pressure gradient, is likely to increase the volume flow rate over static predictions. These considerations may be important in planning long deviated wells where one needs to know, for a given rpm, what maximum borehole length is possible with the pump at hand.

Example 2:   Detailed Spatial Properties Versus "r"

Our computational algorithm does more than calculate annular volume flow rate and average axial speed. This section includes the entire output file from a typical run, in this case "East Greenbriar No. 2", with annotated comments. The input menu is nearly identical to the summary shown in Table 3-1.

Table 3-1
Summary of Input Parameters

O Drill pipe outer radius (inches) = 2.5000
O Borehole radius (inches) = 5.0000
O Axial pressure gradient (psi/ft) = 0.0039
O Drillstring rotation rate (rpm) = 300.0000
O Drillstring rotation rate (rad/sec) = 31.4159
O Fluid exponent "n" (nondimensional) = 0.7240
O Consistency factor (lbf sec^n/sq in) = 0.1861E-04
O Mass density of fluid (lbf^2sec^4/ft) = 1.9000
   (e.g., about 1.9 for water)
O Number of radial "grid" positions = 18

Because the numerical results are based on analytical closed form results, there are no computational inputs; the grid reference in Table 3-1 is a print control parameter. The excessively high value of rpm was chosen for presentation purposes only; the results highlight the qualitative effects of rotation. At the present, the volume flow rate is the only quantity computed numerically; a second-order scheme is applied to our $v_z(r)$'s. All inputs are in "plain English" and are easily understandable. Outputs are similarly "user friendly". All output quantities are defined, along with units, in a printout that precedes tabulated and plotted results. This printout is duplicated in Table 3-2.

Table 3-2
Analytical (Non-Iterative) Solutions Tabulated
versus "r", Nomenclature and Units

r	Annular radial position	(in)
V_z	Velocity in axial z direction	(in/sec)
V_ϑ	Circumferential velocity	(in/sec)
$d\vartheta/dt$ or W .	ϑ velocity	(rad/sec)
	(Note: 1 rad/sec = 9.5493 rpm)	
dV_z/dr	Velocity gradient	(1/sec)
dV_ϑ/dr	Velocity gradient	(1/sec)
dW/dr	Angular speed gradient	(1/(sec x in))
$S_{r\vartheta}$	$r\vartheta$ stress component	(psi)
S_{rz}	rz stress component	(psi)
S_{max}	Sqrt $(S_{rz}^{**}2 + S_{r\vartheta}^{**}2)$	(psi)
dP/dr	Radial pressure gradient	(psi/in)
App-Vis ...	Apparent viscosity	(lbf sec /sq in)
Dissip	Dissipation function	(lbf/(sec x sq in))
Atan V_ϑ/V_z	Angle betw V_ϑ and V_z vectors ..	(deg)
Net Spd	Sqrt $(V_z^{**}2 + V_\vartheta^{**}2)$	(in/sec)
$D_{r\vartheta}$	$r\vartheta$ deformation tensor component	(1/sec)
D_{rz}	rz deformation tensor component..	(1/sec)

The defined quantities are first tabulated, as shown in Table 3-3, as a function of the radial position "r".

Table 3-3
Calculated Quantities vs "r"

r	V_z	V_ϑ	W	$d(V_z)/dr$	$d(V_\vartheta)/dr$	dW/dr
5.00	.601E-04	.279E-04	.559E-05	-.610E+02	-.293E+02	-.586E+01
4.86	.848E+01	.407E+01	.837E+00	-.534E+02	-.293E+02	-.620E+01
4.72	.164E+02	.814E+01	.172E+01	-.460E+02	-.294E+02	-.659E+01
4.58	.237E+02	.122E+02	.266E+01	-.390E+02	-.295E+02	-.702E+01
4.44	.304E+02	.163E+02	.366E+01	-.321E+02	-.297E+02	-.751E+01
4.31	.365E+02	.203E+02	.472E+01	-.256E+02	-.302E+02	-.811E+01
4.17	.418E+02	.244E+02	.585E+01	-.193E+02	-.310E+02	-.885E+01
4.03	.462E+02	.284E+02	.705E+01	-.131E+02	-.324E+02	-.980E+01
3.89	.497E+02	.325E+02	.835E+01	-.672E+01	-.345E+02	-.110E+02
3.75	.521E+02	.366E+02	.975E+01	.273E-04	-.374E+02	-.126E+02
3.61	.533E+02	.407E+02	.113E+02	.738E+01	-.412E+02	-.145E+02
3.47	.532E+02	.449E+02	.129E+02	.157E+02	-.461E+02	-.170E+02
3.33	.516E+02	.492E+02	.148E+02	.251E+02	-.521E+02	-.201E+02
3.19	.483E+02	.536E+02	.168E+02	.358E+02	-.594E+02	-.239E+02
3.06	.432E+02	.582E+02	.190E+02	.480E+02	-.682E+02	-.286E+02
2.92	.361E+02	.630E+02	.216E+02	.619E+02	-.787E+02	-.344E+02
2.78	.266E+02	.680E+02	.245E+02	.778E+02	-.914E+02	-.417E+02
2.64	.147E+02	.732E+02	.277E+02	.959E+02	-.107E+03	-.510E+02
2.50	.000E+00	.785E+02	.314E+02	.117E+03	-.126E+03	-.628E+02

Table 3-3
Calculated Quantities vs "r"   (continued)

r	$S_{r\vartheta}$	S_{rz}	S_{max}	dP/dr	App-Vis	Dissip
5.00	.170E-03	-.355E-03	.393E-03	.143E-13	.582E-05	.266E-01
4.86	.180E-03	-.319E-03	.366E-03	.312E-03	.598E-05	.225E-01
4.72	.191E-03	-.283E-03	.341E-03	.128E-02	.614E-05	.190E-01
4.58	.203E-03	-.246E-03	.318E-03	.298E-02	.630E-05	.161E-01
4.44	.216E-03	-.208E-03	.299E-03	.545E-02	.646E-05	.139E-01
4.31	.230E-03	-.168E-03	.285E-03	.878E-02	.658E-05	.123E-01
4.17	.245E-03	-.128E-03	.277E-03	.131E-01	.665E-05	.115E-01
4.03	.262E-03	-.869E-04	.277E-03	.184E-01	.665E-05	.115E-01
3.89	.282E-03	-.442E-04	.285E-03	.248E-01	.658E-05	.124E-01
3.75	.303E-03	.175E-09	.303E-03	.326E-01	.643E-05	.143E-01
3.61	.327E-03	.459E-04	.330E-03	.420E-01	.622E-05	.175E-01
3.47	.353E-03	.936E-04	.365E-03	.532E-01	.598E-05	.223E-01
3.33	.383E-03	.144E-03	.409E-03	.665E-01	.573E-05	.292E-01
3.19	.417E-03	.196E-03	.461E-03	.824E-01	.547E-05	.388E-01
3.06	.456E-03	.251E-03	.520E-03	.102E+00	.523E-05	.518E-01
2.92	.501E-03	.309E-03	.588E-03	.125E+00	.499E-05	.693E-01
2.78	.552E-03	.370E-03	.665E-03	.152E+00	.476E-05	.928E-01
2.64	.611E-03	.436E-03	.751E-03	.186E+00	.455E-05	.124E+00
2.50	.681E-03	.507E-03	.849E-03	.226E+00	.434E-05	.166E+00

Table 3-3
Calculated Quantities vs "r" (continued)

r	V_z	V_ϑ	Atan V_ϑ/V_z	Net Spd	$D_{r\vartheta}$	D_{rz}
5.00	.601E-04	.279E-04	.249E+02	.663E-04	.146E+02	-.305E+02
4.86	.848E+01	.407E+01	.256E+02	.940E+01	.151E+02	-.267E+02
4.72	.164E+02	.814E+01	.264E+02	.183E+02	.155E+02	-.230E+02
4.58	.237E+02	.122E+02	.272E+02	.267E+02	.161E+02	-.195E+02
4.44	.304E+02	.163E+02	.281E+02	.345E+02	.167E+02	-.161E+02
4.31	.365E+02	.203E+02	.291E+02	.417E+02	.175E+02	-.128E+02
4.17	.418E+02	.244E+02	.303E+02	.483E+02	.184E+02	-.965E+01
4.03	.462E+02	.284E+02	.316E+02	.542E+02	.197E+02	-.653E+01
3.89	.497E+02	.325E+02	.332E+02	.593E+02	.214E+02	-.336E+01
3.75	.521E+02	.366E+02	.351E+02	.636E+02	.236E+02	.137E-04
3.61	.533E+02	.407E+02	.374E+02	.670E+02	.262E+02	.369E+01
3.47	.532E+02	.449E+02	.402E+02	.696E+02	.295E+02	.783E+01
3.33	.516E+02	.492E+02	.436E+02	.713E+02	.334E+02	.125E+02
3.19	.483E+02	.536E+02	.480E+02	.722E+02	.381E+02	.179E+02
3.06	.432E+02	.582E+02	.534E+02	.725E+02	.436E+02	.240E+02
2.92	.361E+02	.630E+02	.602E+02	.726E+02	.502E+02	.309E+02
2.78	.266E+02	.680E+02	.686E+02	.730E+02	.579E+02	.389E+02
2.64	.147E+02	.732E+02	.786E+02	.746E+02	.673E+02	.480E+02
2.50	.000E+00	.785E+02	.900E+02	.785E+02	.785E+02	.584E+02

At this point, the total volume flow rate is computed and presented in textual form, that is,

Total volume flow rate (cubic in/sec) = .2026E+04

(gal/min) = .5261E+03

A run-time screen menu prompts the user with regard to quantities he would like displayed in ASCII file plots. These plots produce tabulated numerical values as well as visual trend information versus "r". The complete list of quantities was given previously. Plots corresponding to "East Greenbriar No. 2" are shown next with annotations.

Axial speed $V_z(r)$:

```
   r                    0

 5.00    .6014E-04    |
 4.86    .8478E+01    | *
 4.72    .1639E+02    |   *           No-slip
 4.58    .2373E+02    |     *         conditions
 4.44    .3044E+02    |       *       enforced
 4.31    .3647E+02    |         *
 4.17    .4175E+02    |           *
 4.03    .4618E+02    |            *
 3.89    .4966E+02    |             *
 3.75    .5207E+02    |              *
 3.61    .5329E+02    |               *
 3.47    .5317E+02    |              *
 3.33    .5157E+02    |               *
 3.19    .4830E+02    |             *
 3.06    .4320E+02    |           *
 2.92    .3606E+02    |         *
 2.78    .2665E+02    |       *
 2.64    .1472E+02    |    *
 2.50    .0000E+00    |
```

Figure 3-2. Plotted Results

Circumferential speed $V_{\vartheta}(r)$:

r		0

```
5.00    .2793E-04    I
4.86    .4069E+01    *
4.72    .8138E+01    I *
4.58    .1220E+02    I *
4.44    .1626E+02    I  *
4.31    .2031E+02    I   *
4.17    .2436E+02    I    *
4.03    .2841E+02    I     *      Maximum speed is
3.89    .3247E+02    I      *     at drillstring
3.75    .3655E+02    I       *
3.61    .4068E+02    I        *
3.47    .4488E+02    I         *
3.33    .4918E+02    I         *
3.19    .5361E+02    I          *
3.06    .5820E+02    I           *
2.92    .6298E+02    I            *
2.78    .6797E+02    I             *
2.64    .7316E+02    I              *
2.50    .7854E+02    I                *
```

Figure 3-3.   Plotted Results (continued)

Angular speed W(r):

r	0	

r		
5.00	.5586E-05	I
4.86	.8370E+00	I
4.72	.1723E+01	*
4.58	.2662E+01	I*
4.44	.3659E+01	I *
4.31	.4718E+01	I * Maximum speed is
4.17	.5847E+01	I * at drillstring
4.03	.7053E+01	I *
3.89	.8349E+01	I *
3.75	.9748E+01	I *
3.61	.1127E+02	I *
3.47	.1293E+02	I *
3.33	.1475E+02	I *
3.19	.1678E+02	I *
3.06	.1905E+02	I *
2.92	.2159E+02	I *
2.78	.2447E+02	I *
2.64	.2772E+02	I *
2.50	.3142E+02	I *

Figure 3-4. Plotted Results (continued)

Velocity gradient d(V$_z$)/dr (r):

r			
		0	
5.00	-.6096E+02	*		
4.86	-.5339E+02	*		
4.72	-.4604E+02	*		
4.58	-.3896E+02	*		Consistent with
4.44	-.3215E+02	*		axial velocity
4.31	-.2561E+02	*		solution
4.17	-.1930E+02	*		
4.03	-.1307E+02	*|	
3.89	-.6724E+01	*		
3.75	.2730E-04			
3.61	.7377E+01			
3.47	.1566E+02		*	
3.33	.2505E+02		*	
3.19	.3575E+02		*	
3.06	.4796E+02		*	
2.92	.6188E+02		*	
2.78	.7776E+02		*	
2.64	.9593E+02		*	
2.50	.1168E+03		*	

Figure 3-5.   Plotted Results (continued)

Velocity gradient d(V_ϑ)/dr (r):

r   0

5.00	-.2929E+02	*	
4.86	-.2932E+02	*	
4.72	-.2938E+02	*	
4.58	-.2949E+02	*	
4.44	-.2974E+02	*	
4.31	-.3021E+02	*	
4.17	-.3104E+02	*	
4.03	-.3240E+02	*	
3.89	-.3447E+02	*	
3.75	-.3738E+02	*	
3.61	-.4123E+02	*	
3.47	-.4612E+02	*	
3.33	-.5214E+02	*	
3.19	-.5944E+02	*	
3.06	-.6821E+02	*	
2.92	-.7874E+02	*	
2.78	-.9142E+02	*	
2.64	-.1068E+03	*	
2.50	-.1257E+03		

Figure 3-6.   Plotted Results (continued)

Angular speed gradient dW/dr (r):

r		0
5.00	-.5857E+01	* |
4.86	-.6204E+01	* |
4.72	-.6586E+01	* |
4.58	-.7016E+01	* |
4.44	-.7514E+01	* |
4.31	-.8112E+01	* |
4.17	-.8853E+01	* |
4.03	-.9796E+01	* |
3.89	-.1101E+02	* |
3.75	-.1257E+02	* |
3.61	-.1454E+02	* |
3.47	-.1701E+02	* |
3.33	-.2007E+02	* |
3.19	-.2386E+02	* |
3.06	-.2856E+02	* |
2.92	-.3440E+02	* |
2.78	-.4172E+02	* |
2.64	-.5098E+02	* |
2.50	-.6283E+02	|

Figure 3-7. Plotted Results (continued)

Stress component $S_{r\vartheta}$ (r):

r	0		
5.00	.1703E-03	I	*
4.86	.1802E-03	I	*
4.72	.1910E-03	I	*
4.58	.2027E-03	I	*
4.44	.2156E-03	I	*
4.31	.2297E-03	I	*
4.17	.2453E-03	I	*
4.03	.2625E-03	I	*
3.89	.2816E-03	I	*
3.75	.3028E-03	I	*
3.61	.3266E-03	I	*
3.47	.3532E-03	I	*
3.33	.3832E-03	I	*
3.19	.4173E-03	I	*
3.06	.4561E-03	I	*
2.92	.5006E-03	I	*
2.78	.5519E-03	I	*
2.64	.6115E-03	I	*
2.50	.6813E-03	I	*

Figure 3-8. Plotted Results (continued)

Stress component S$_{rz}$ (r):

r			0
5.00	-.3546E-03	*		
4.86	-.3190E-03	*		
4.72	-.2827E-03	*		
4.58	-.2456E-03	*		
4.44	-.2075E-03	*		
4.31	-.1685E-03	*		
4.17	-.1283E-03	*		
4.03	-.8694E-04	*		
3.89	-.4422E-04	*	
3.75	.1754E-09		
3.61	.4589E-04		*
3.47	.9365E-04		*
3.33	.1435E-03		*
3.19	.1958E-03		*
3.06	.2507E-03		*
2.92	.3087E-03		*
2.78	.3703E-03		*
2.64	.4360E-03		*
2.50	.5065E-03		*

Figure 3-9. Plotted Results (continued)

Maximum stress S$_{max}$ (r):

r		0

```
5.00    .3933E-03   I        *
4.86    .3664E-03   I       *
4.72    .3412E-03   I        *        This stress may be
4.58    .3184E-03   I       *         responsible for
4.44    .2992E-03   I      *          erosion of borehole
4.31    .2849E-03   I      *          wall and cuttings
4.17    .2768E-03   I     *           beds
4.03    .2765E-03   I     *
3.89    .2850E-03   I     *
3.75    .3028E-03   I     *
3.61    .3298E-03   I      *
3.47    .3654E-03   I       *
3.33    .4092E-03   I         *
3.19    .4609E-03   I           *
3.06    .5205E-03   I            *
2.92    .5881E-03   I             *
2.78    .6646E-03   I              *
2.64    .7510E-03   I                *
2.50    .8490E-03   I                  *
```

Figure 3-10.   Plotted Results (continued)

Radial pressure gradient dP/dr (r):

r	0		
5.00	.1430E-13	I	
4.86	.3121E-03	I	
4.72	.1285E-02	I	
4.58	.2977E-02	I	
4.44	.5452E-02	I	Partial
4.31	.8781E-02	*	centrifugal
4.17	.1305E-01	*	effects
4.03	.1836E-01	I*	(see Equation 3-29)
3.89	.2484E-01	I *	
3.75	.3265E-01	I *	
3.61	.4200E-01	I *	
3.47	.5316E-01	I *	
3.33	.6649E-01	I *	
3.19	.8244E-01	I *	
3.06	.1016E+00	I *	
2.92	.1246E+00	I *	
2.78	.1524E+00	I *	
2.64	.1858E+00	I *	
2.50	.2261E+00	I *	

Figure 3-11. Plotted Results (continued)

Apparent viscosity vs "r":

r		0
5.00	.5816E-05	I *
4.86	.5976E-05	I *
4.72	.6140E-05	I *
4.58	.6304E-05	I *
4.44	.6455E-05	I *
4.31	.6577E-05	I *
4.17	.6650E-05	I *
4.03	.6652E-05	I *
3.89	.6576E-05	I *
3.75	.6426E-05	I *
3.61	.6220E-05	I *
3.47	.5982E-05	I *
3.33	.5729E-05	I *
3.19	.5475E-05	I *
3.06	.5227E-05	I *
2.92	.4989E-05	I *
2.78	.4762E-05	I *
2.64	.4545E-05	I * Varies
2.50	.4338E-05	I * with "r" !

Figure 3-12.   Plotted Results (continued)

Dissipation function vs "r":

r	0	
5.00	.2660E-01	| *
4.86	.2247E-01	| *
4.72	.1896E-01	| * The greatest heat is
4.58	.1609E-01	|* produced near the
4.44	.1387E-01	|* drillstring surface
4.31	.1234E-01	|*
4.17	.1152E-01	|*
4.03	.1149E-01	|*
3.89	.1235E-01	|*
3.75	.1427E-01	|*
3.61	.1748E-01	| *
3.47	.2232E-01	| *
3.33	.2923E-01	| *
3.19	.3881E-01	| *
3.06	.5182E-01	| *
2.92	.6933E-01	| *
2.78	.9276E-01	| *
2.64	.1241E+00	| *
2.50	.1662E+00	| *

Figure 3-13.   Plotted Results (continued)

Angle between V_{ϑ} and V_z vectors, Atan V_{ϑ}/V_z (r):

r   0

5.00	.2491E+02	|	*
4.86	.2564E+02	|	*
4.72	.2640E+02	|	* This angle measures
4.58	.2721E+02	|	* extent of helical
4.44	.2811E+02	|	* annular flow in
4.31	.2911E+02	|	* degrees
4.17	.3026E+02	|	*
4.03	.3160E+02	|	*
3.89	.3318E+02	|	*
3.75	.3507E+02	|	*
3.61	.3736E+02	|	*
3.47	.4017E+02	|	*
3.33	.4364E+02	|	*
3.19	.4798E+02	|	*
3.06	.5341E+02	|	*
2.92	.6021E+02	|	*
2.78	.6859E+02	|	*
2.64	.7862E+02	|	*
2.50	.9000E+02	|	*

Figure 3-14.   Plotted Results (continued)

Magnitude of total speed vs r:

r   0

```
5.00    .6631E-04    |
4.86    .9404E+01    | *
4.72    .1830E+02    |    *
4.58    .2668E+02    |       *
4.44    .3451E+02    |         *
4.31    .4175E+02    |           *
4.17    .4834E+02    |             *
4.03    .5422E+02    |              *
3.89    .5933E+02    |               *
3.75    .6362E+02    |                *
3.61    .6704E+02    |                 *
3.47    .6958E+02    |                 *
3.33    .7126E+02    |                  *
3.19    .7216E+02    |                  *
3.06    .7249E+02    |                  *
2.92    .7257E+02    |                  *
2.78    .7300E+02    |                  *
2.64    .7462E+02    |                   *
2.50    .7854E+02    |                      *
```

Figure 3-15. Plotted Results (continued)

Deformation tensor element $D_{r\vartheta}$ (r):

r	0	
5.00	.1464E+02	| *
4.86	.1508E+02	| *
4.72	.1555E+02	| *
4.58	.1608E+02	| *
4.44	.1670E+02	| *
4.31	.1746E+02	| *
4.17	.1844E+02	| *
4.03	.1973E+02	| *
3.89	.2141E+02	| *
3.75	.2356E+02	| *
3.61	.2625E+02	| *
3.47	.2952E+02	| *
3.33	.3345E+02	| *
3.19	.3811E+02	| *
3.06	.4363E+02	| *
2.92	.5016E+02	| *
2.78	.5795E+02	| *
2.64	.6727E+02	| *
2.50	.7854E+02	| *

Figure 3-16.   Plotted Results (continued)

Deformation tensor element D_{rz} (r):

```
  r                          0
      _____
 5.00   -.3048E+02        *    |
 4.86   -.2669E+02        *    |
 4.72   -.2302E+02         *   |
 4.58   -.1948E+02         *   |
 4.44   -.1607E+02          *  |
 4.31   -.1281E+02           * |
 4.17   -.9648E+01            *|
 4.03   -.6535E+01            *|
 3.89   -.3362E+01             *
 3.75    .1365E-04             |
 3.61    .3689E+01             |
 3.47    .7828E+01             | *
 3.33    .1253E+02             |  *
 3.19    .1788E+02             |   *
 3.06    .2398E+02             |    *
 2.92    .3094E+02             |     *
 2.78    .3888E+02             |       *
 2.64    .4796E+02             |         *
 2.50    .5839E+02             |             *
```

Figure 3-17.   Plotted Results (continued)

Finally, the computer algorithm calculates radially averaged quantities using the definition

$$F_{avg} = \int_{R_P}^{R_B} F(r)\ dr\ /(R_B - R_P) \tag{3-83}$$

and a second-order accurate integration scheme. Note that this is not a volume weighted average. When properties vary rapidly over r, the linear average (or *any average*) may not be meaningful as a correlation or analysis parameter. Table 3-4 displays computed averages.

Table 3-4

Averaged Values of Annular Quantities

Average V_z (in/sec)   = .3512E+02

       (ft/min)   = .1756E+03

Average V_ϑ (in/sec)   = .3737E+02

Average W (rad/sec)   = .1160E+02

Average total speed (in/sec) = .5379E+02

Average angle between V_z and V_ϑ (deg) = .4189E+02

Average $d(V_z)/dr$ (1/sec) = .0000E+00

Average $d(V_\vartheta)/dr$ (1/sec) = -.5028E+02

Average dW/dr (1/(sec X in)) = -.1906E+02

Average dP/dr (psi/in) = .5718E-01

Average $S_{r\vartheta}$ (psi) = .3410E-03

Average S_{rz} (psi) = .2432E-04

Average S_{max} (psi) = .4146E-03

Average dissipation function (lbf/(sec sq in)) = .3753E-01

Average apparent viscosity (lbf sec/sq in) = .5876E-05

Average $D_{r\vartheta}$ (1/sec) = .3094E+02

Average D_{rz} (1/sec) = .4445E+01

Example 3:   Another Calculation

We repeated the calculations for "East Greenbriar No. 2" with all parameters unchanged except for the fluid exponent, which we increased to a near-Newtonian level of 0.9 (again, 1.0 is the Newtonian value). In the first run, we considered a static, nonrotating drillstring with a "rpm" of 0.001, and obtained a volume flow rate of 196.2 gal/min. This differs from our earlier 373.6 gal/min, which assumed a fluid exponent of n = 0.724. That is, a 24% increase in the fluid exponent n resulted in a 47% decrease in flow rate; these numbers show how sensitive results are to changes in n. The axial speeds, apparent viscosities and averaged parameter values obtained are given in Figures 3-18 and 3-19, and Table 3-5. Note how the apparent viscosity is almost constant everywhere with respect to radial position; the well known, localized "pinch" is found near the center of the annulus, where the axial velocity gradient vanishes.

Axial speed Vz(r):

```
    r                   0

  5.00    .8649E-05    |
  4.86    .2948E+01    | *
  4.72    .6058E+01    |    *
  4.58    .9010E+01    |       *
  4.44    .1171E+02    |          *
  4.31    .1410E+02    |             *
  4.17    .1613E+02    |               *
  4.03    .1776E+02    |                 *
  3.89    .1897E+02    |                  *
  3.75    .1972E+02    |                   *
  3.61    .1998E+02    |                    *
  3.47    .1971E+02    |                   *
  3.33    .1889E+02    |                  *
  3.19    .1747E+02    |                 *
  3.06    .1541E+02    |              *
  2.92    .1268E+02    |           *
  2.78    .9239E+01    |       *
  2.64    .5029E+01    |   *
  2.50    .0000E+00    |
```

Figure 3-18.   Plotted Results

Apparent viscosity vs "r":

r	0		
5.00	.1341E-04	I	*
4.86	.1357E-04	I	*
4.72	.1375E-04	I	*
4.58	.1397E-04	I	*
4.44	.1424E-04	I	*
4.31	.1457E-04	I	*
4.17	.1502E-04	I	*
4.03	.1568E-04	I	*
3.89	.1690E-04	I	*
3.75	.4468E-04	I	*
3.61	.1683E-04	I	*
3.47	.1555E-04	I	*
3.33	.1483E-04	I	*
3.19	.1433E-04	I	*
3.06	.1394E-04	I	*
2.92	.1362E-04	I	*
2.78	.1335E-04	I	*
2.64	.1311E-04	I	*
2.50	.1289E-04	I	*

Figure 3-19.   Plotted Results (continued)

For our second run, we retain the foregoing parameters with the exception of drillstring rpm, which we increase significantly for test purposes from 0.001 to 300 (the fluid exponent is still 0.9). The volume flow rate computed was 232.9 gpm, higher than the 196.2 gpm obtained above by 18.7%. Thus, even for "almost Newtonian" power law fluids, the effect of rotation allows a higher flow rate for the same pressure drop. Thus, to produce the lower flow rate, a pump having less pressure output "than normal" would suffice. Computed results are shown in Figures 3-20 and 3-21 and Table 3-6.

Table 3-5
Averaged Values of Annular Quantities

Average V_z (in/sec) $= .1305E+02$
(ft/min) $= .6523E+02$
Average V_ϑ (in/sec) $= .3535E-03$
Average W (rad/sec) $= .1074E-03$
Average total speed (in/sec) $= .1305E+02$
Average angle between V_z and V_ϑ (deg) $= .2441E+01$
Average $d(V_z)/dr$ (1/sec) $= .0000E+00$
Average $d(V_\vartheta)/dr$ (1/sec) $= -.4288E-03$
Average dW/dr (1/(sec X in)) $= -.1630E-03$
Average dP/dr (psi/in) $= .4719E-11$
Average $S_{r\vartheta}$ (psi) $= .7903E-08$
Average S_{rz} (psi) $= .2432E-04$
Average S_{max} (psi) $= .2088E-03$
Average dissipation function (lbf/(sec sq in)) $= .4442E-02$
Average apparent viscosity (lbf sec/sq in) $= .1617E-04$
Average $D_{r\vartheta}$ (1/sec) $= .2681E-03$
Average D_{rz} (1/sec) $= .9912E+00$

Axial speed Vz(r):

```
   r                    0
                        _____
 5.00    .3053E-04    |
 4.86    .4256E+01    |  *
 4.72    .8125E+01    |      *
 4.58    .1159E+02    |          *
 4.44    .1465E+02    |             *
 4.31    .1727E+02    |                *
 4.17    .1943E+02    |                   *
 4.03    .2113E+02    |                     *
 3.89    .2232E+02    |                      *
 3.75    .2300E+02    |                       *
 3.61    .2312E+02    |                        *
 3.47    .2267E+02    |                       *
 3.33    .2160E+02    |                      *
 3.19    .1988E+02    |                    *
 3.06    .1748E+02    |                  *
 2.92    .1434E+02    |              *
 2.78    .1041E+02    |          *
 2.64    .5656E+01    |     *
 2.50    .0000E+00    |
```

Figure 3-20.   Plotted Results

Apparent viscosity vs "r":

r   0

5.00	.1294E-04		*
4.86	.1298E-04		*
4.72	.1300E-04		*
4.58	.1300E-04		*
4.44	.1299E-04		*
4.31	.1296E-04		*
4.17	.1291E-04		*
4.03	.1284E-04		*
3.89	.1276E-04		*
3.75	.1266E-04		*
3.61	.1255E-04		*
3.47	.1243E-04		*
3.33	.1231E-04		*
3.19	.1218E-04		*
3.06	.1205E-04		*
2.92	.1191E-04		*
2.78	.1177E-04		*
2.64	.1163E-04		*
2.50	.1148E-04		*

Figure 3-21.   Plotted Results (continued)

Table 3-6
Averaged Values of Annular Quantities

Average V_z (in/sec)   = .1539E+02

(ft/min)   = .7693E+02

Average V_ϑ (in/sec)   = .3730E+02

Average W (rad/sec)   = .1162E+02

Average total speed (in/sec) = .4150E+02

Average angle between V_z and V_ϑ (deg) = .5993E+02

Average $d(V_z)/dr$ (1/sec) = .0000E+00

Average $d(V_\vartheta)/dr$ (1/sec) = -.4303E+02

Average dW/dr (1/(sec X in)) = -.1654E+02

Average dP/dr (psi/in) = .5811E-01

Average $S_{r\vartheta}$ (psi) = .6717E-03

Average S_{rz} (psi) = .2432E-04

Average S_{max} (psi) = .7149E-03

Average dissipation function (lbf/(sec sq in)) = .4851E-01

Average apparent viscosity (lbf sec/sq in) = .1251E-04

Average $D_{r\vartheta}$ (1/sec) = .2733E+02

Average D_{rz} (1/sec) = .1350E+01

In the foregoing example, the effect of increased drillstring rpm has increased the average borehole maximum stress by 3.42 times; this may be of interest to wellbore stability. The apparent viscosity in this example, unlike the previous, is nearly constant everywhere and does not "pinch out".

Closing Remarks. The analytical solutions derived in this chapter are of fundamental rheological interest. They are also useful in drilling and production applications, as we will see in Chapter 5. They allow us to study various "what if" questions quickly and efficiently. These solutions also provide a simple means to correlate experimental data nondimensionally, but they are restricted to narrow annular passages.

REFERENCES

---

1. Schlichting, H., *Boundary Layer Theory*, New York: McGraw-Hill, 1968.
2. Slattery, J.C., *Momentum, Energy, and Mass Transfer in Continua*, New York: Robert E. Krieger Publishing Company, 1981.
3. Savins, J.G., and Wallick, G.C., "Viscosity Profiles, Discharge Rates, Pressures, and Torques for a Rheologically Complex Fluid in a Helical Flow," *A.I.Ch.E. Journal*, Vol. 12, No. 2, March 1966, pp. 357-363.
4. Fredrickson, A.G., and Bird, R.B., "Non-Newtonian Flow in Annuli," *Ind. Eng. Chem.*, 1958, Vol. 50, p. 347.

4

Recirculating Annular Flows

Problems with cuttings accumulation, flow blockage and resultant stuck pipe in deviated wells are becoming increasingly important operational issues as interest in horizontal drilling continues. For small angles (ß) from the vertical, annular flows and hole cleaning are well understood. But beyond thirty degrees, these issues are rife with challenging questions; many unexplained, confusing and conflicting observations are reported by different investigators.

Chapter 5 addresses several problems related to cuttings transport using the eccentric and rotating flow models developed earlier for homogeneous fluids. Cuttings accumulation is dangerous because the resulting blockage of the annular space increases the possibility of stuck pipe.

In Chapters 2 and 3, the pressure field was assumed to be uniform across the annulus; the velocity field is therefore unidirectional, with the fluid flowing axially from high pressure regions to low. These assumptions are reasonable since numerous flows do behave in this manner.

But not always. It turns out that *dangerous flow blockage can arise from fluid-dynamical effects alone*, even without the presence of cuttings or debris in the annulus. This is made possible by the combined effects of flow heterogeneity and drillpipe (or casing) inclination. The existence of cuttings, of course, only increases the severity of the problem. A special class of annular flows lends itself to such strange occurrences; these "recirculating vortex flows" are studied in this chapter. These fluid masses are not unlike "stationary tornados" sitting in a windstorm, entraining any debris convected by the outer mainstream flow.

WHAT ARE RECIRCULATING VORTEX FLOWS?

In deviated holes where circulation has been temporarily interrupted, weighting material such as barite, drilled cuttings or cement additives may fall out of suspension. This gravity segregation has mass density increasing downwards. And this stable stratification, which we collectively refer to as "barite sag", is thought to be responsible for the trapped, self-contained "recirculation zones" or "bubbles" observed by many experimenters.

These bubbles contain rotating, swirling, "ferris-wheel-like" motions within their interiors; the external fluid which flows around them "sees" these zones as stationary obstacles that impede their axial movement up the annulus. Excellent color video tapes showing these vortex-like motions in detail have been produced by M-I Drilling Fluids, and may be viewed at its laboratory facilities in Houston upon request.

These strange occurrences are just that; their appearances seem to be sporadic and unpredictable, as much myth as reality. However, once they are formed, they remain as stable fluid-dynamical structures that are extremely difficult to remove. They are dangerous and undesirable because of their tendency to entrain cuttings, block axial flow, and increase the possibility of stuck pipe. Why do these bubbles form? What are the controlling parameters? How can their occurrences be prevented?

Detailed study of M-I's tapes suggests that the recirculating flows form independently of viscosity and rheology to leading order. They appear to be primarily inertia dominated, while nonconservative effects play only a minor role in sustaining or damping the motion. This leaves the component of density stratification normal to the hole axis as the primary culprit; it alone is responsible for the highly three-dimensional pressure field that drives local pockets of secondary flow. It is possible, of course, to have multiple bubbles coexisting along a long deviated hole.

Again these recirculating bubbles, observed near pipe bends, stabilizers and other obstructions, are important for various practical reasons. First, they block the streamwise axial flow, resulting in the need for increased pressures to pass a given volume flow rate. Second, because they entrain the mud and further trap drilled cuttings, they are a likely cause of stuck pipe. Third, the external flow modified by these bubbles can also affect the very process of cuttings bed formation and removal itself.

Fortunately, these bubbles can be studied, modelled and characterized in a rather simple manner. This chapter identifies the *nondimensional* channel parameter **Ch** responsible for vortex bubble formation and describes the physics of these recirculating flows. The equations of motion are given and solved using finite difference methods for several practical flows. The detailed bubble development process is described and illustrated in a sequence of computer generated pictures showing streamline evolution.

MOTIVATING IDEAS AND CONTROLLING VARIABLES

The governing momentum equations used are the steady Euler equations, which describe large amplitude, inviscid shear flow in both stratified and unstratified media (Schlichting, 1968; Turner, 1973). For the problem at hand, they simplify to

$$\rho \{uu_x + vu_y\} = - p_x \tag{4-1}$$

$$\rho \{uv_x + vv_y\} = - p_y - g \rho \cos \alpha \tag{4-2}$$

$$\{\rho u\}_x + \{\rho v\}_y = 0 \tag{4-3}$$

Equations 4-1 and 4-2 are momentum equations in the x and y directions, while Equation 4-3 describes mass conservation. In obtaining Equations 4-1 to 4-3, we have assumed that the vortex flow is contained in a two-dimensional rectangular box in the plane of the borehole axis (x) and the direction of density stratification (y). This is based on experimental observation: the vortical flows do *not* wrap around the drillpipe. In the above equations, u and v are velocities in the x and y directions, respectively.

Subscripts indicate partial derivatives. Also, ρ is a fluid density which varies linearly with y far upstream, p(x,y) is the unknown pressure field, g is the acceleration due to gravity, and α is the angle the borehole axis makes with the horizontal ($\alpha + \beta = 90^o$; also see Chapters 2 and 3).

Equations 4-1 to 4-3 introduce subtle complications. Since we do not suppress the dependence of p(x,y) on the cross-coordinate y, the streamwise pressure gradient cannot be specified a priori as in Chapters 2 and 3. It must be determined as part of an inverse scheme.

Nondimensional parameters are important to *understanding* physical events. The well known Reynolds number, which measures the relative effects of inertia to viscous forces, is one example of a nondimensional parameter. It alone, for example, dictates the onset of turbulence; also, like Reynolds numbers imply dynamically similar flow patterns. Analogous nondimensional variables are used in different areas of physics; for instance, the mobility ratio in reservoir engineering or the Mach number in high speed aerodynamics.

Close examination of Equations 4-1 to 4-3 using affine transformations shows that the physics of bubble formation depends on a *single* nondimensional variable **Ch** characterizing the channel flow. It is constructed from the combination of two simpler ones. The first is a Froude number $U^2/gL \cos \alpha$, where U is the average oncoming speed and L is the channel height between the pipe and borehole walls. The second is a relative measure of stratification, say $d\rho/\rho_{ref}$ ($d\rho$ might represent the density difference between the bottom and top of the annulus, and ρ_{ref} might be taken as the arithmetic average). The combined parameter **Ch** of practical significance is

$$Ch = U^2 \rho_{ref}/gLd\rho \cos \alpha \qquad (4\text{-}4)$$

We now summarize our numerical findings. For large values of **Ch**, recirculation bubbles will *not* form; the streamlines of motion are essentially straight and the rheology-dominated models developed in Chapters 2 and 3 apply. For small **Ch**'s of order unity, small recirculation zones *do* form, and elongate in the streamwise dimension as **Ch** decreases. For still smaller values, solutions with wavy upstream flows are found, which may or may not be physically realistic.

Equations 4-1 to 4-3 can be solved using "brute force" computational methods, but they are more cleverly treated by introducing the streamfunction used by aerodynamicists and reservoir engineers. When the problem is reformulated in this manner, the result is a nonlinear Poisson equation that can be easily integrated using fast, iterative elliptic solvers.

Streamlines are obtained by connecting computed streamfunction elevations having like values. The arithmetic difference in streamfunction between any two points is directly proportional to the volume flow rate passing through the two points. Velocity and pressure fields can be obtained by post-processing the computed streamfunction solutions.

Typically, the solution obtained for a 20 x 40 mesh will require approximately thirty seconds of computing time on IBM PC-XTs with math co-processors. In Figures 4-1 to 4-6, we have allowed the flow to "disappear" into a "mathematical sink" at the corner (in practice, the distance to a suitable obstacle divided by the height L will appear as a second ratio). This sink simulates the presence of obstacles or pipe elbows located upstream. With decreasing values of **Ch**, the appearance of an elongating recirculation bubble is seen. The computed streamline patterns, again constructed by drawing level streamfunction contours, depict an ever-worsening flowfield (background streamfunction numbers are shown for reference).

The stand-alone vortexes so obtained are inherently stable, since they represent patches of angular momentum that physical laws insist must be conserved. In this sense, they are not unlike isolated trailing aircraft tip vortices, swirling flows that persist indefinitely in the air until dissipation renders them harmless. However, annular bubbles are worse: the channel flow itself is what drives them, perpetuates them, and fuels their ability to do harm.

```
105 105  96  98 102  99  97 101 101  98  99 101  99  98 100 100  98 100
 53  73  82  86  89  91  93  94  95  96  96  97  97  98  98  99  99 100
 32  54  67  74  80  83  86  89  90  92  93  94  95  96  97  98  99 100
 23  41  55  64  71  76  80  83  86  88  90  92  93  95  96  97  98 100
 18  33  46  56  64  70  75  78  82  85  87  89  91  93  95  96  98 100
 14  28  40  50  58  64  69  74  78  81  84  87  89  92  94  96  98 100
 12  24  35  44  52  59  65  70  74  78  82  85  87  90  93  95  97 100
 11  22  32  40  48  55  61  67  71  75  79  83  86  89  92  94  97 100
 10  20  29  37  45  52  58  63  68  73  77  81  84  87  91  94  97 100
  9  18  27  35  42  49  55  61  66  71  75  79  83  86  90  93  96 100
  8  17  25  33  40  46  53  58  64  69  73  77  81  85  89  93  96 100
  8  16  23  31  38  44  51  56  62  67  72  76  80  84  88  92  96 100
  7  15  22  29  36  43  49  55  60  65  70  75  79  83  88  92  96 100
  7  14  21  28  35  41  47  53  59  64  69  74  78  83  87  91  95 100
  7  14  21  27  34  40  46  52  57  63  68  73  77  82  86  91  95 100
  6  13  20  26  33  39  45  51  56  62  67  72  77  81  86  91  95 100
  6  13  19  26  32  38  44  50  55  61  66  71  76  81  86  90  95 100
  6  12  19  25  31  37  43  49  55  59  65  70  75  80  85  90  95 100
  6  12  18  25  31  37  43  48  54  59  65  70  75  80  85  90  95 100
  6  12  18  24  30  36  42  48  53  59  64  69  75  80  85  90  95 100
  6  12  18  24  30  36  41  47  53  58  64  69  74  79  84  89  94 100
  6  12  18  24  29  35  41  47  52  58  63  69  74  79  84  89  94 100
  5  11  17  23  29  35  41  46  52  57  63  68  74  79  84  89  94 100
  5  11  17  23  29  35  40  46  52  57  63  68  73  79  84  89  94 100
  5  11  17  23  29  34  40  46  51  57  62  68  73  78  84  89  94 100
  5  11  17  23  28  34  40  45  51  57  62  67  73  78  84  89  94 100
  5  11  17  23  28  34  40  45  51  56  62  67  73  78  83  89  94 100
  5  11  17  22  28  34  39  45  51  56  62  67  73  78  83  89  94 100
  5  11  17  22  28  34  39  45  50  56  62  67  72  78  83  89  94 100
  5  11  17  22  28  34  39  45  50  56  61  67  72  78  83  89  94 100
  5  11  17  22  28  33  39  45  50  56  61  67  72  78  83  89  94 100
  5  11  16  22  28  33  39  45  50  56  61  67  72  78  83  89  94 100
  5  11  16  22  28  33  39  44  50  56  61  67  72  78  83  89  94 100
  5  11  16  22  28  33  39  44  50  56  61  67  72  78  83  89  94 100
  5  11  16  22  28  33  39  44  50  55  61  67  72  78  83  89  94 100
  5  11  16  22  28  33  39  44  50  55  61  66  72  78  83  89  94 100
  5  11  16  22  27  33  39  44  50  55  61  66  72  77  83  88  94 100
  5  11  16  22  27  33  39  44  50  55  61  66  72  77  83  88  94 100
  5  11  16  22  27  33  39  44  50  55  61  66  72  77  83  88  94 100
  5  11  16  22  27  33  39  44  50  55  61  66  72  77  83  88  94 100
```

Figure 4-1. Channel Flow, Ch = 1.0.

0	105	105	96	98	102	99	97	101	101	98	99	101	99	98	100	100	98	100
0	54	74	83	88	91	93	94	96	96	97	98	98	98	99	99	99	99	100
0	33	55	68	77	82	86	89	91	93	94	95	96	97	98	98	99	99	100
0	24	43	57	67	74	80	84	87	89	91	93	94	96	97	97	98	99	100
0	19	35	49	59	67	74	79	83	86	88	91	92	94	95	97	98	99	100
0	15	30	43	53	61	68	74	79	82	86	88	90	92	94	96	97	98	100
0	13	26	38	48	56	64	70	75	79	83	86	88	91	93	95	96	98	100
0	12	23	34	44	52	60	66	71	76	80	83	86	89	92	94	96	98	100
0	11	21	31	41	49	56	63	68	73	77	81	85	88	90	93	95	97	100
0	10	20	29	38	46	53	60	65	71	75	79	83	86	89	92	95	97	100
0	9	18	27	36	43	51	57	63	68	73	77	81	85	88	91	94	97	100
0	8	17	26	34	41	48	55	61	66	71	76	80	83	87	90	93	97	100
0	8	16	24	32	40	46	53	59	64	69	74	78	82	86	90	93	96	100
0	8	16	23	31	38	45	51	57	63	68	73	77	81	85	89	93	96	100
0	7	15	22	30	37	43	50	56	61	66	71	76	80	84	88	92	96	100
0	7	14	22	29	36	42	48	54	60	65	70	75	79	84	88	92	96	100
0	7	14	21	28	35	41	47	53	59	64	69	74	78	83	87	91	95	100
0	7	14	20	27	34	40	46	52	58	63	68	73	78	82	87	91	95	100
0	6	13	20	26	33	39	45	51	57	62	67	72	77	82	86	91	95	100
0	6	13	19	26	32	38	44	50	56	61	67	72	76	81	86	90	95	100
0	6	13	19	25	32	38	44	50	55	61	66	71	76	81	86	90	95	100
0	6	12	19	25	31	37	43	49	55	60	66	70	75	80	85	90	95	100
0	6	12	18	25	31	37	43	48	54	59	65	70	75	80	85	90	95	100
0	6	12	18	24	30	36	42	48	53	59	64	69	74	79	85	89	95	100
0	6	12	18	24	30	36	41	47	53	58	64	69	74	79	84	89	94	100
0	6	12	18	23	29	35	41	47	52	58	63	69	74	79	84	89	94	100
0	5	11	17	23	29	35	41	46	52	57	63	68	74	79	84	89	94	100
0	5	11	17	23	29	35	40	46	52	57	63	68	73	79	84	89	94	100
0	5	11	17	23	29	34	46	51	57	62	68	73	78	84	89	94	100	
0	5	11	17	23	29	34	46	51	57	62	68	73	78	84	89	94	100	
0	5	11	17	23	28	34	40	45	51	57	62	67	73	78	83	89	94	100
0	5	11	17	23	28	34	40	45	51	56	62	67	73	78	83	89	94	100
0	5	11	17	22	28	34	39	45	51	56	62	67	73	78	83	89	94	100
0	5	11	16	22	28	34	39	45	50	56	61	67	72	78	83	89	94	100
0	5	11	16	22	28	33	39	45	50	56	61	67	72	78	83	89	94	100
0	5	11	16	22	28	39	44	50	56	61	67	72	78	83	89	94	100	

Figure 4-2. Channel Flow, Ch = 0.5.

105	105	96	98	102	99	97	101	101	98	99	101	99	98	100	100	98	100
55	76	85	90	94	96	98	99	100	100	100	101	101	100	100	100	100	100
35	58	72	81	87	92	95	97	99	100	101	101	101	101	101	100	100	100
25	47	62	73	81	87	92	95	97	99	100	101	101	101	101	101	100	100
21	39	55	66	76	83	88	92	95	98	99	100	101	101	101	101	100	100
18	34	49	61	71	79	85	90	93	96	98	100	100	101	101	100	100	100
16	31	45	57	67	75	82	87	91	94	97	99	100	100	100	100	100	100
14	28	41	53	63	71	78	84	89	93	95	97	99	99	100	100	100	100
13	26	39	50	60	68	76	82	87	91	94	96	98	99	99	100	100	100
12	25	37	47	57	66	73	79	85	89	92	95	96	98	98	99	99	100
12	28	35	45	55	63	71	77	82	87	90	93	95	97	98	99	99	100
11	22	33	43	53	61	68	75	80	85	89	92	94	96	97	98	99	100
11	21	32	42	51	59	66	73	78	83	87	90	93	95	96	98	99	100
10	21	31	40	49	57	64	71	77	82	86	89	92	94	96	97	98	100
10	20	30	39	47	55	63	69	75	80	84	88	91	93	95	97	98	100
9	19	29	38	46	54	61	68	73	78	83	86	90	92	94	96	98	100
9	19	28	36	45	53	60	66	72	77	81	85	89	91	94	96	98	100
9	18	27	35	44	51	58	65	70	76	80	84	88	90	93	95	97	100
9	17	26	35	43	50	57	63	69	74	79	83	87	90	92	95	97	100
8	17	26	34	42	49	56	62	68	73	78	82	86	89	92	95	97	100
8	17	25	33	41	48	55	61	67	72	77	81	85	88	91	94	97	100
8	16	24	32	40	47	54	60	66	71	76	80	84	88	91	94	97	100
8	16	24	32	39	46	53	59	65	70	75	79	83	87	90	94	97	100
8	15	23	31	38	45	52	58	64	69	74	78	83	86	90	93	96	100
7	15	23	30	37	44	51	57	63	68	73	78	82	86	89	93	96	100
7	15	22	30	37	44	50	56	62	67	72	77	81	85	89	93	96	100
7	15	22	29	36	43	49	55	61	67	72	76	81	85	89	92	96	100
7	14	22	29	36	42	49	55	60	66	71	76	80	84	88	92	96	100
7	14	21	28	35	42	48	54	60	65	70	75	80	84	88	92	96	100
7	14	21	28	35	41	47	53	59	64	70	74	79	83	88	92	96	100
7	14	21	27	34	41	47	53	58	64	69	74	79	83	87	91	96	100
7	13	20	27	34	40	46	52	58	63	69	73	78	83	87	91	95	100
6	13	20	27	33	40	46	52	57	63	68	73	78	82	87	91	95	100
6	13	20	26	33	39	45	51	57	62	68	73	77	82	87	91	95	100
6	13	20	26	33	39	45	51	56	62	67	72	77	82	86	91	95	100
6	13	19	26	32	38	44	50	56	61	67	72	77	81	86	91	95	100
6	13	19	26	32	38	44	50	56	61	66	71	76	81	86	90	95	100
6	13	19	25	32	38	44	50	55	61	66	71	76	81	86	90	95	100
6	12	19	25	31	37	43	49	55	60	66	71	76	81	85	90	95	100
6	12	19	25	31	37	43	49	54	60	65	70	76	80	85	90	95	100
6	12	19	25	31	37	43	49	54	60	65	70	75	80	85	90	95	100

Figure 4-3. Channel Flow, Ch = 0.35.

105	105	96	98	102	99	97	101	101	98	99	101	99	98	100	100	98	100
55	77	87	93	96	99	100	102	102	103	103	103	103	102	102	101	100	100
36	60	75	85	92	97	100	103	104	105	106	106	105	104	103	102	101	100
27	50	67	79	88	95	100	103	106	107	108	108	107	106	105	103	101	100
23	43	60	74	84	92	98	103	106	108	109	109	109	108	106	104	102	100
20	39	56	70	81	90	97	102	106	109	110	110	110	109	107	105	102	100
18	36	52	66	78	88	96	102	106	109	111	111	111	109	108	105	102	100
17	34	50	64	76	86	94	101	105	109	111	111	111	110	108	105	103	100
16	33	48	62	74	84	93	100	105	108	111	112	111	110	108	106	103	100
16	32	47	60	72	83	92	99	104	108	110	112	111	110	108	106	103	100
15	31	45	59	71	82	90	98	103	108	110	111	111	110	108	106	103	100
15	30	44	58	70	80	89	97	103	107	110	111	111	110	108	106	103	100
15	29	44	57	69	79	88	96	102	106	109	111	111	110	108	106	103	100
14	29	43	56	68	78	87	95	101	106	109	110	110	110	108	106	103	100
14	28	42	55	67	77	87	94	100	105	108	110	110	109	108	105	103	100
14	28	42	54	66	77	86	93	99	104	107	109	110	109	107	105	102	100
14	28	41	54	65	76	85	93	99	103	107	108	109	109	107	105	102	100
14	27	41	53	65	75	84	92	98	103	106	108	109	108	107	105	102	100
13	27	40	53	64	74	83	91	97	102	105	107	108	108	106	105	102	100
13	27	40	52	63	74	83	90	96	101	105	107	107	107	106	104	102	100
13	26	39	52	63	73	82	89	96	100	104	106	107	107	106	104	102	100
13	26	39	51	62	72	81	89	95	100	103	105	106	106	105	104	102	100
13	26	39	51	62	72	80	88	94	99	103	105	106	106	105	104	102	100
13	26	38	50	61	71	80	87	93	98	102	104	105	105	105	103	102	100
13	25	38	50	60	70	79	87	93	98	101	104	105	105	104	103	101	100
12	25	37	49	60	70	78	86	92	97	101	103	104	105	104	103	101	100
12	25	37	49	59	69	78	85	91	96	100	102	104	104	104	103	101	100
12	25	37	48	59	69	77	85	91	96	99	102	103	104	103	102	101	100
12	24	36	48	58	68	76	84	89	94	98	101	102	103	102	102	101	100
12	24	36	47	58	67	76	83	89	94	98	101	102	103	102	102	101	100
12	24	36	47	57	67	75	83	89	94	97	100	102	102	102	102	101	100
12	24	35	47	57	66	75	82	88	93	97	100	102	102	102	101	101	100
12	24	35	46	56	66	74	81	88	92	96	99	101	102	102	101	101	100
12	23	35	46	56	65	74	81	87	92	96	99	100	101	101	100	100	100
11	23	35	45	55	65	73	80	86	91	95	98	100	101	101	101	100	100
11	23	34	45	55	64	72	80	86	91	95	98	99	101	101	101	100	100
11	23	34	45	55	64	72	79	85	90	94	97	99	100	101	101	100	100
11	23	34	44	54	63	71	78	85	90	94	97	99	100	100	100	100	100
11	22	33	44	54	63	71	78	84	89	93	96	98	99	100	100	100	100
11	22	33	44	53	62	70	77	83	88	93	96	98	99	100	100	100	100
11	22	33	43	53	62	70	77	83	88	92	95	97	99	100	100	100	100

Figure 4-4. Channel Flow, Ch = 0.320.

```
0  105 105  96  98 102  99  97 101 101  98  99 101  99  98 100 100  98 100
0   55  77  87  93  97  99 101 102 103 103 103 103 103 102 102 101 100 100
0   36  61  76  86  93  98 101 104 105 106 106 106 106 105 104 102 101 100
0   27  50  67  80  89  96 101 104 107 108 109 109 108 107 105 104 102 100
0   23  44  61  75  85  93 100 104 107 109 110 111 110 109 107 105 102 100
0   20  40  57  71  82  91  99 104 108 110 112 112 111 110 108 105 102 100
0   19  37  53  68  80  90  97 104 108 111 112 113 112 111 109 106 103 100
0   18  35  51  65  78  88  96 103 108 111 113 114 113 111 109 106 103 100
0   17  34  49  63  76  87  95 102 108 111 113 114 113 112 110 107 103 100
0   16  33  48  62  75  85  94 102 107 111 113 114 114 112 110 107 103 100
0   16  32  47  61  73  84  94 101 107 111 113 114 114 112 110 107 103 100
0   16  31  46  60  72  83  93 100 106 110 113 114 114 112 110 107 103 100
0   15  31  45  59  72  83  92 100 106 110 113 114 114 112 110 107 103 100
0   15  30  45  58  71  82  91  99 105 110 112 114 113 112 110 107 103 100
0   15  30  44  58  70  81  91  98 105 109 112 113 113 112 110 107 103 100
0   15  30  44  57  70  81  90  98 104 109 111 113 113 112 110 107 103 100
0   15  29  44  57  69  80  89  97 103 108 111 113 113 112 109 107 103 100
0   14  29  43  56  69  79  89  97 103 108 111 112 112 111 109 106 103 100
0   14  29  43  56  68  79  88  96 102 107 110 112 112 111 109 106 103 100
0   14  29  42  56  68  78  88  96 102 107 110 111 112 111 109 106 103 100
0   14  28  42  55  67  78  87  95 101 106 109 111 111 110 109 106 103 100
0   14  28  42  55  67  77  87  94 101 105 109 110 111 110 108 106 103 100
0   14  28  42  54  66  77  86  94 100 105 108 110 111 110 108 106 103 100
0   14  28  41  54  66  76  86  93 100 104 108 110 110 109 108 106 103 100
0   14  27  41  53  65  76  85  92  99 103 107 109 109 109 107 105 102 100
0   14  27  41  53  65  75  84  92  98 103 106 108 109 109 107 105 102 100
0   13  27  40  53  64  75  84  91  98 102 106 108 109 108 107 105 102 100
0   13  27  40  52  64  74  83  91  97 102 105 107 108 108 108 106 105 102 100
0   13  27  40  52  63  73  82  90  97 101 104 107 107 107 106 104 102 100
0   13  26  39  51  63  73  82  89  96 101 104 106 107 107 106 104 102 100
0   13  26  39  51  62  72  81  89  95 100 103 105 105 106 105 104 102 100
0   13  26  39  51  62  72  81  88  94  99 102 105 106 106 105 104 102 100
0   13  26  38  50  61  71  80  88  94  99 102 105 106 106 105 104 102 100
0   13  26  38  50  61  71  80  87  93  98 102 104 105 105 105 103 102 100
0   13  25  38  50  61  70  79  87  93  98 101 104 105 105 105 103 101 100
0   13  25  38  49  60  70  79  86  93  97 101 103 105 105 104 103 101 100
0   12  25  37  49  60  70  78  86  92  97 101 103 104 105 104 103 101 100
0   12  25  37  49  60  69  78  85  92  97 100 103 104 104 104 103 101 100
```

Figure 4-5. Channel Flow, Ch = 0.319.

```
105 105  96  98 102  99  97 101 101  98  99 101  99  98 100 100  98 100
 56  77  87  93  97 100 101 103 103 103 104 104 104 103 103 102 101 100 100
 36  61  76  87  94  99 102 105 106 107 108 107 107 106 104 103 101 100
 28  51  68  81  90  97 102 106 108 110 110 110 109 108 106 104 102 100
 23  45  62  76  87  95 102 107 110 112 113 113 112 110 108 105 102 100
 21  41  58  72  84  94 101 107 111 113 114 114 113 112 109 106 103 100
 19  38  55  70  82  92 101 107 111 114 116 116 115 113 110 107 103 100
 18  36  53  68  81  91 100 107 112 115 117 117 117 114 111 108 104 100
 18  35  52  66  79  90 100 107 112 115 117 118 117 115 112 108 104 100
 17  34  50  65  78  90  99 107 112 116 118 118 117 115 112 108 104 100
 17  34  50  64  78  89  99 106 112 116 118 119 118 116 113 109 104 100
 17  33  49  64  77  89  98 106 112 116 119 119 118 116 113 109 104 100
 16  33  49  63  77  88  98 106 112 116 119 119 116 116 113 109 104 100
 16  33  48  63  76  88  98 106 112 116 119 120 119 117 113 109 105 100
 16  33  48  63  76  88  98 106 112 116 119 120 119 117 114 109 105 100
 16  32  48  62  76  87  97 106 112 116 119 120 119 117 114 109 105 100
 16  32  48  62  76  87  97 106 112 116 119 120 119 117 114 109 105 100
 16  32  48  62  75  87  97 105 112 116 119 120 119 117 114 110 105 100
 16  32  48  62  75  87  97 105 112 116 119 120 119 117 114 110 105 100
 16  32  47  62  75  87  97 105 112 116 119 120 119 117 114 110 105 100
 16  32  47  62  75  87  97 105 111 116 119 120 119 117 114 110 105 100
 16  32  47  62  75  86  97 105 111 116 119 120 119 117 114 110 105 100
 16  32  47  61  75  86  96 105 111 116 119 120 119 117 114 109 105 100
 16  32  47  61  75  86  96 105 111 116 119 120 119 117 114 109 105 100
 16  32  47  61  75  86  96 105 111 116 118 119 119 117 114 109 105 100
 16  32  47  61  74  86  96 105 111 116 118 119 119 117 114 109 105 100
 16  32  47  61  74  86  96 104 111 116 118 119 119 117 114 109 105 100
 16  32  47  61  74  86  96 104 111 115 118 119 119 117 113 109 105 100
 16  32  47  61  74  86  96 104 111 115 118 119 119 117 113 109 105 100
 16  32  47  61  74  86  96 104 111 115 118 119 119 117 113 109 105 100
 16  31  47  61  74  86  96 104 110 115 118 119 118 116 113 109 105 100
 16  31  47  61  74  86  96 104 110 115 118 119 118 116 113 109 105 100
 16  31  47  61  74  86  96 104 110 115 118 119 118 116 113 109 105 100
 16  31  47  61  74  85  95 104 110 115 118 119 118 116 113 109 104 100
 16  31  46  61  74  85  95 104 110 115 118 119 118 116 113 109 104 100
 16  31  46  61  74  85  95 104 110 115 118 119 118 116 113 109 104 100
```

Figure 4-6. Channel Flow, Ch = 0.318.

HOW TO AVOID STAGNANT BUBBLES

We have shown that recirculating zones can develop from interactions between inertia and gravity forces. These bubbles form when density stratification, hole deviation and pump rate fulfill certain special conditions. These conditions are elegantly captured in a single channel variable, the nondimensional parameter $Ch = U^2 \rho_{ref}/gLd\rho \cos \alpha$. Moreover, the resulting flowfields can be efficiently computed and displayed, thus allowing us to better understand their dynamical consequences.

Suppressing recirculating flows is easily accomplished: *avoid small values of the nondimensional* Ch *parameter*. Small values, as is evident from Equation 4-4, can result from different isolated effects. For example, Ch decreases as the hole becomes more horizontal, as density differences become more pronounced, or as pumping rates decrease. But none of these factors alone controls the physics; it is the combination *taken together* that controls bubble formation and perhaps the fate of a drilling program.

We have modelled the problem as the single-phase flow of a continuous stratified fluid, rather than as the combined motion of a dual-phase fluid and solid continuum. This simplifies the mathematical issues without sacrificing the essential physical details. For practical purposes, the parameter Ch can be viewed as a "danger indicator" signalling impending cuttings transport or stuck pipe problems. It is the single most important parameter whenever interrupted circulation or poor suspension properties leads to gravity segregation and settling of weighting materials in drilling mud.

These same considerations also apply to cementing, where density segregation due to gravity and *slow velocities* are both likely. When recirculation zones form in either the mud or the cement above or beneath the casing, the displacement effectiveness of the cement is severely impeded. The result is mud left in place, an undesirable one necessitating remedial squeeze jobs.

We emphasize that the vortical bubbles considered here are *not* the Taylor vortices studied in the classical fluid mechanics of homogeneous flows. Taylor vortices are "doughnuts" which would normally wrap around, in our case, the pipe; to the author's knowledge, they have not been observed in drilling applications. Taylor vortices owe their existence to finite drillstring length effects and represent the result of different physical mechanisms.

A practical example. We have discussed the dynamical significance of the nondimensional parameter Ch that appears in the normalized equations of motion. For use in practical estimates, the channel variable may be written more clearly as

$$Ch = U^2 \rho_{ref}/gLd\rho \cos \alpha \qquad (4\text{-}4)$$

that is, as a multiplicative sequence of nondimensional entities

$$\mathbf{Ch} = (U^2/gL) \times (\rho_{ref}/d\rho) \times (1/\cos \alpha) \tag{4-5}$$

Let us consider an annular flow studied in the cuttings transport examples of Chapter 5. For the 2 and 5 inch pipe and borehole radii, the cross-sectional area is $\pi(5^2 - 2^2)$ or 66 in^2.

The experimental data used in Discussions 1 and 2 of Chapter 5 assume oncoming linear velocities of 1.91, 2.86 and 3.82 ft/sec. Since 1 ft/sec corresponds to a volume flow rate of 1 ft/sec x 66 in^2 or 205.7 gpm, the flow rates there are 393, 588 and 786 gpm. So at the lowest flow rate of 393 gpm - a reasonable field number - the average linear speed over the entire annulus is approximately 2 ft/sec. But the low-side average will be much smaller, say 0.5 ft/sec. And if the pipe is displaced halfway down, the length scale L will be roughly (5"-2")/2 or 0.13 ft.

Thus, the first factor in Equation 4-5 takes the value $U^2/gL = (0.5)^2/(32.2 \times 0.13) = 0.06$. If we assume a 20% density stratification, then $\rho_{ref}/d\rho = 5.0$; the product of the two factors is 0.30. For a highly deviated well inclined 70^0 from the vertical axis, $\alpha = 90^0 - 70^0 = 20^0$ and cos 20^0 = 0.94. Thus, $\mathbf{Ch} = 0.30/0.94 = 0.32$. This value, as Figures 4-1 to 4-6 show, lies just at the threshold of danger. Velocities lower than the assumed value are even more likely to sustain recirculatory flows; higher ones, in contrast, are safer.

Of course, the numbers used above are only estimates; a three-dimensional, viscous solution is required to establish true length and velocity scales. But these "ballpark" results clearly show that bottomhole conditions typical of those used in drilling and cementing *are* associated with low values of $\mathbf{Ch}$ near unity.

We emphasize again that $\mathbf{Ch}$ is the only nondimensional parameter appearing in Equations 4-1 to 4-3. Another one describing the geometry of the annular domain would normally appear through boundary conditions. For convenience though - and for the sake of argument only - we have replaced this requirement with an idealized "sink". In any real calculation, exact geometrical effects must be included to complete the formulation.

Also note that our recirculating flows get "worse" as the borehole becomes more horizontal; that is, $\mathbf{Ch}$ decreases as a becomes smaller. This is in stark contrast to the unidirectional homogeneous flows of Chapter 5 which, as we will prove, perform worst near 45^0, at least with respect to cuttings transport efficiency. The structure of Equation 4-4 correctly shows that in near-vertical wells with α approaching 90^0, $\mathbf{Ch}$ tends towards infinity; thus, the effects of flow blockage due to the vortical bubbles considered here are relegated to highly deviated wells only.

Again, flow properties such as local velocity, viscous stress, shear rate and pressure can be obtained from the computed streamfunction straightforwardly. They may be useful correlation parameters for cuttings transport efficiency and local bed buildup. Continuing research is underway, exploring similarities between this problem and the density-dependent flows studied in dynamic meteorology and oceanography.

REFERENCES

---

1. Schlichting, H., *Boundary Layer Theory,* New York: McGraw-Hill, 1968.
2. Turner, J.S., *Buoyancy Effects in Fluids*, London: Cambridge University Press, 1973.

5

Applications to
Drilling and Production

In Chapters 2, 3 and 4, we formulated and solved three distinct annular flow models and gave numerous calculated results. These models remove many of the restrictions usually made regarding eccentricity, rotation and flow heterogeneity. The methods also produce fast and stable solutions, at the same time displaying results without any investment in graphical tools. The computer software can be used with minimal training in petroleum engineering and computer modeling.

This chapter, the central focus of this book, deals with practical applications in drilling and production. Topics discussed include cuttings transport in deviated holes, stuck pipe, cementing and coiled tubing return flow analysis. For the casual reader, brief summaries of the three models are offered first.

RECAPITULATION: MODEL SUMMARIES

The first annular flow model applies to homogeneous Newtonian, power law, Bingham plastic and Herschel-Bulkley fluids. It assumes a nonrotating drillpipe (or casing) in an inclined hole, but no restrictions are placed on the annular geometry itself. The analysis handles eccentric circular drillpipes and boreholes. But it also models, with no extra difficulty or computational expense, deformed borehole walls due to erosion, swelling or cuttings bed buildup; and, as well, even square drill collars with stabilizers. A brief description is offered in Chin (1990a).

The exact PDEs are solved on boundary conforming grid systems which yield high resolution in tight spaces. They handle geometry exactly, so that slot flow and bipolar coordinate approximations are unnecessary. The unconditionally stable finite difference program, requiring ten seconds per simulation on a "386" machine, has been successfully executed over two thousand times without diverging.

To help understand the lengthy output, which includes annular velocity, apparent viscosity, two components each of viscous stress and shear rate, Stokes product and dissipation function, a special character-based graphics program was developed. This utility overlays results on the annular geometry itself, thus facilitating physical interpretation and visual correlation of computed quantities with annular position. The algorithms and graphics software are written in standard Fortran; they are easily ported to most hardware platforms with minimal modification. Calculations and displays for several difficult geometries are given in Chapter 2.

The second flow model provides approximate analytical solutions for concentric annular flows containing rotating drillstrings or casings. Here, a "narrow annulus" assumption is invoked; hole inclination, as in the first model, scales out by introducing a normalized pressure. Closed form solutions are derived for axial and circumferential velocity, pressure, viscous stress, deformation rate, apparent viscosity and local heat generation. These results are restricted to Newtonian and power law fluids; all the properties listed are given as explicit functions of the radial coordinate "r". Again, built-in, portable graphical utilities permit quick and convenient displays of all relevant flow parameters.

The third and final model considers density heterogeneities due to gravity stratification. This opens up the possibility of reversed flow in the axial direction, and annular blockage of an entirely fluid-dynamical nature. But analysis shows that the existence of "trapped bubbles" depends on a single nondimensional parameter only; presumably, these flows may be easy to control and avoid.

CUTTINGS TRANSPORT IN DEVIATED WELLS

Recent industry interest in horizontal and highly deviated wells has heightened the importance of annular flow modeling as it relates to hole cleaning. Cuttings transport to the surface is generally impeded by virtue of hole orientation; this is worsened by decreased "low-side" annular velocities due to pipe eccentricity. In addition, the blockage created by bed buildup decreases overall flow rate, further reducing cleaning efficiency. In what could possibly be a self-sustaining, destabilizing process, stuck pipe is a likely end result. This section discloses new cuttings transport correlations and suggests simple predictive measures to avoid bed buildup. Good hole cleaning and bed removal, of course, are important to cementing as well.

Few useful annular flow models are available despite their practical importance. The nonlinear equations governing power law viscous fluids, for example, must be solved with difficult no-slip velocity conditions for highly eccentric geometries. Recent slot flow models offer some improvement over parallel plate approaches. Still, because they unrealistically require slow radial variations in the circumferential direction, large errors are possible. Even when they apply, these models can be cumbersome; they involve "elliptic integrals" which are too awkward for field use. Bipolar coordinate models accurately simulate eccentric flows with circular pipes and boreholes; however, they cannot be extended to applications containing washouts and cuttings beds.

In this section, the eccentric flow model is used to interpret field and laboratory results. Because the model actually simulates reality, it has been possible to correlate problems associated with cuttings transport and stuck pipe to unique average mechanical properties of the computed flowfield. These correlations are discussed next.

Discussion 1:   Water Base Muds

Detailed computations using the eccentric model are described, assuming a power law fluid, which correspond to the comprehensive suite of cuttings transport experiments recently conducted at the University of Tulsa (Becker, Azar and Okrajni, 1989). For a fixed inclination and oncoming flow rate, we importantly demonstrate that "cuttings concentration" correlates *linearly* with *the mean viscous shear stress averaged over the lower half of the annulus*. This result was first reported in Chin (1990b).

Thus impending cuttings problems can be eased by first determining the existing average stress level; and then, adjusting n, k and gpm values to increase that stress. Physical arguments supporting our correlations will be given. We emphasize that the present approach is completely predictive and deterministic; it does *not* require empirical assumptions related to the "equivalent hydraulic radius" or to any "pipe to annulus conversion factors".

Detailed experimental results for cuttings concentration, a useful indicator of transport efficiency and carrying capacity, were obtained at the University of Tulsa's large scale flow loop. Fifteen bentonite-polymer, water-based muds, for three average flow rates (1.91, 2.86 and 3.82 ft/sec), at three borehole inclinations from vertical (30, 45 and 70 deg), were tested. Table 1 of Becker et al. (1989) summarizes all measured mud properties, along with specific power law exponents n and consistency factors k. We emphasize that "water base" does *not* imply Newtonian flow; in fact, the reported values of n differ substantially from unity. The annular geometry consisted of a 2" radius pipe, displaced downward by 1.5" in a 5" radius borehole; also, the pipe rotated at 50 rpm.

With flow rate and hole inclination fixed, the authors cross-plot the nondimensional cuttings concentration C versus particular rheological properties for each mud type used. These included apparent viscosity, plastic viscosity (PV), yield point (YP), YP/PV, initial and ten-minute gel strength, "effective viscosity", k, and Fann dial readings at various rpms. Typically, the correlations obtained were poor, with one exception to be discussed.

That good correlations were not possible, of course, is not surprising; the "fluid properties" in Becker et al. (1989) are rotational viscometer readings describing the test instrument only. That is, they have no real bearing to the actual annular geometry and its downhole flow.

These cross-plots and tables, numbering over twenty, were nevertheless studied in detail; using them, the entire laboratory database was reconstructed. The concentric, rotating flow program described in Chapter 3 was run to show that rpm effects were likely to be insignificant. The eccentric annular model of Chapter 2 was then executed for each of the 135 experimental points; detailed results for *calculated* apparent viscosity, shear rate, viscous stress and axial velocity, all of which varied spatially, were tabulated and statistically analyzed along with the experimental data.

Numerous cross-plots were produced, examined and interpreted. The most meaningful correlation parameter found was the *mean viscous shear stress*, obtained by averaging computed values over the bottom half of the annulus, where cuttings in directional wells are known to form beds.

Figures 5-1, 5-2 and 5-3 display the cuttings concentration versus our mean shear stress for different average flow speeds and inclination angles ß from the vertical. Each plotted symbol represents a distinct test mud. Calculated correlation coefficients averaged a high 0.91 value. Our correlations apply to laminar flow only; the flattened portions of the curves refer to turbulent conditions. Computationally, the latter can be simulated with a simple subroutine change, but this would require a suitable turbulence model.

The computer program produces easily understood information. Figure 5-4 displays, for example, calculated areal results for viscous shear stress in the highly visual format described earlier. Tabulated results, in this case for "Mud No. 10" at 1.91 ft/sec, show that the "24" at the bottom refers to "0.00024 psi" (thus the numbers in the plot, when multiplied by 10^{-5}, give the actual psi level). A high value of "83" is seen on the upper pipe surface; lows are generally obtained away from solid surfaces and at the annular floor. The average of these calculated values, taken over the bottom half of the annulus, supply the mean stress points on the horizontal axes of Figures 5-1, 5-2 and 5-3.

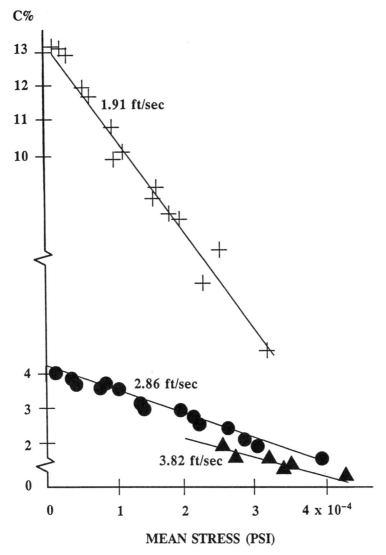

Figure 5-1. Cuttings Transport Correlation, $\beta = 30$ deg.

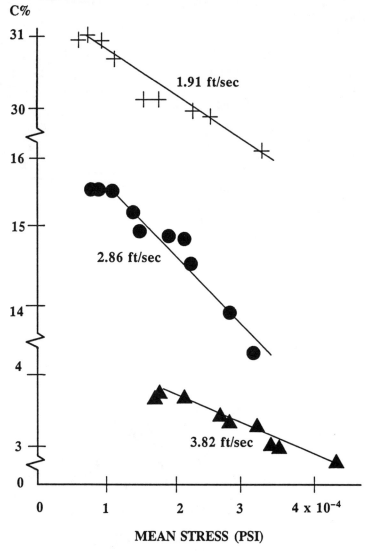

Figure 5-2. Cuttings Transport Correlation, β = 45 deg.

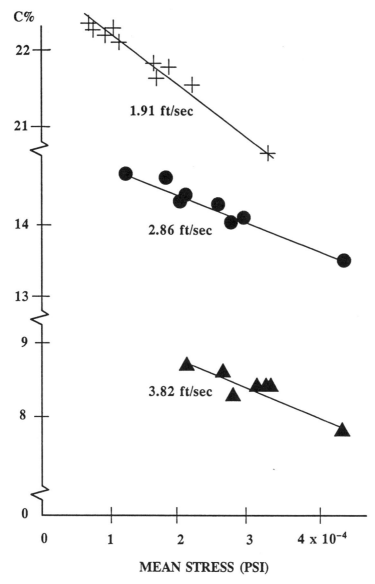

Figure 5-3. Cuttings Transport Correlation, $\beta = 70$ deg.

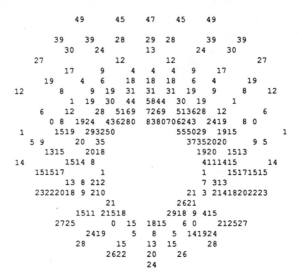

Figure 5-4. Viscous Stress.

Becker et al. (1989) noted that the best data fit, obtained through trial and error, was obtained with low shear rate parameters; in particular, Fann dial (stress) readings at low rotary speeds like 6 rpm. This corresponds to a shear rate of 10/sec. Our exact, computed results gave averaged rates of 7-9/sec for *all* the mud samples at 1.91 ft/sec; similarly, 11-14/sec at 2.86 ft/sec, and 14-19/sec at 3.82 ft/sec. Since these are in the 10/sec range, they explain why a 6 rpm correlation worked, at least in their particular test setup.

But in general, the authors' "low rpm" recommendation will *not* apply a priori; each nonlinear annular flow presents a unique physical problem with its own characteristic shears. In general, pipe to hole diameter ratio as well as eccentricity enter the equation; these can increase the low shear rates obtained above by orders of magnitude. But this poses no difficulty since downhole properties *can* be obtained with minimal effort.

Cuttings removal in near-vertical holes with ß < 10⁰ is well understood; cleaning efficiency is proportional to annular velocity, or more precisely, the "Stokes product" between relative velocity and local viscosity. For inclined wells, the usual notions regarding unimpeded settling velocities do not apply because different physical processes are at work. Cuttings travel almost immediately to the low side of the annulus, a consequence of gravity segregation; they remain there and form beds that may or may not slide downward.

These truss or lattice-like structures have well defined mechanical yield stresses; the right amount of viscous friction will erode the cuttings bed, the same way mud circulation limits dynamic filter cake growth. This explains our success in using bottom-averaged viscous stress as the correlation parameter. The straight line fit also indicates that bed properties are linear in an elastic sense. These ideas, of course, are not entirely new. Slurry pipeline designers, for example, routinely consider "boundary shear" and "critical tractive force". They have successfully modeled sediment beds as "series of superposed layers" with distinct yield strengths (Streeter, 1961). However, these studies are restricted to Newtonian carrier fluids in circular conduits.

While viscous shear emerges as the dominant transport parameter, its role was by no mean obvious at the outset. Other correlation quantities tested include vertical and lateral components of shear rates and stresses, axial velocity, apparent viscosity and the Stokes product. These correlated somewhat well, particularly at low inclinations, but shear stress almost *always* worked. Take apparent viscosity, for example. Whereas Figure 6 of Becker et al. (1989) shows significant wide-band scatter, citing *rotational viscometer* values ranging from 1 to 50 cp, our *exact* computations gave good correlations with *actual* apparent viscosities ranging up to 400 cp. Computed viscosities expectedly showed no meaningful connection to the apparent viscosities given by the University of Tulsa investigators, because the latter were inferred from unrealistic Fann dial readings. This point is illustrated quantitatively later.

We emphasize that Figures 5-1, 5-2 and 5-3 are based on unweighted muds. The effect of "pure increases in fluid density" should not alter computed shear stresses, at least theoretically, since the convective terms in the governing equations vanish for straight holes. In practice, however, oilfield weighting materials are likely to alter n and k; thus, some change in stress might be anticipated. The effects of buoyancy, of course, will help without regard to changes in shear.

We have shown how cuttings concentration correlates in a satisfactory manner with *the mean viscous shear stress averaged over the lower half of the annulus.* Thus impending hole cleaning problems can be alleviated by first determining the existing average stress level; and then, by adjusting n, k and gpm values in the actual drilling fluid to increase that stress. Once this danger zone is past, additives can be used to reduce shear stress and hence mud pump pressure requirements. Simply increasing gpm may also help, although the effect of rheology on stress is probably more significant.

Interestingly, Seeberger et al. (1989) described an important field study where extremely high velocities together with very high yield points did not alleviate hole cleaning problems. They suggested that extrapolated YP values may not be useful indicators of transport efficiency. Also, the authors pointed to the importance of elevated stress levels at low shear rates in cleaning large diameter holes at high angles. They experimentally showed how oil and water base muds having like rheograms, despite their obvious textural or "look and feel" differences, will clean with like efficiencies. This implies that a knowledge of n and k alone suffices in characterizing real muds.

The procedure suggested above requires minimal change to field operations. Standard viscometer readings, plotted on "log-log" paper or used in handbook formulas, still represent required information. But they should only be used to determine actual downhole properties insofar as providing n and k for detailed computer analysis. YP and PV, arising from older Bingham models, play no direct role in the present methodology although these parameters sometimes offer useful correlations. Also, results obtained from the eccentric flow model should be available in tabulated form, and preferably be accessible as software at the drilling site.

Discussion 2:   Cuttings Transport Database

The viscometer properties and cuttings concentrations data for the fifteen muds (at all angles and flow rates), together with exact computed results for shear rate, stress, apparent viscosity, annular speed and Stokes product have been assembled into a comparative database for continuing study. These detailed results are available from the author.

Table 5-1
Bottom Averaged Fluid Properties @ 1.91 ft/sec

Mud	n	k lbf sec^n /in^2	Shear Rate 1/sec	Shear Stress (psi)	Apparent-Viscosity (cp)	
1	1.00	0.15E-6	9.1	0.13E-5	1	(1)
2	0.74	0.72E-5	8.1	0.29E-4	27	(8)
3	0.59	0.13E-4	7.8	0.34E-4	35	(5)
4	0.74	0.14E-4	8.3	0.59E-4	54	(15)
5	0.59	0.25E-4	7.6	0.67E-4	71	(9)
6	0.42	0.57E-4	7.4	0.95E-4	116	(6)
7	0.74	0.24E-4	8.1	0.97E-4	89	(25)
8	0.59	0.43E-4	7.6	0.11E-3	118	(15)
9	0.42	0.94E-4	7.5	0.16E-3	191	(10)
10	0.74	0.38E-4	8.2	0.16E-3	143	(40)
11	0.59	0.68E-4	7.7	0.18E-3	190	(24)
12	0.42	0.15E-3	7.5	0.25E-3	307	(16)
13	0.74	0.48E-4	8.0	0.19E-3	180	(50)
14	0.59	0.85E-4	7.6	0.22E-3	237	(30)
15	0.42	0.19E-3	7.4	0.32E-3	388	(20)

Tables 5-1, 5-2 and 5-3 summarize typical bottom-averaged results for the eccentric hole used in the University of Tulsa experiments. Computed results show that the bottom of the hole supports a low shear rate flow, ranging from 10 to 20 reciprocal seconds, approximately. These values are consistent with the authors' low shear rate conclusions, established by trial and error from the experimental data.

However, their rule of thumb is not universally correct; for example, the same muds and flow rates gave high shear rate results for several different downhole geometries. Shear rates *can* vary substantially depending on eccentricity and diameter ratio. Direct analysis is the only legitimate and final arbiter.

Table 5-2
Bottom Averaged Fluid Properties @ 2.86 ft/sec

Mud	n	k lbf sec^n /in^2	Shear Rate 1/sec	Shear Stress (psi)	Apparent-Viscosity (cp)	
1	1.00	0.15E-6	14	0.20E-5	1	(1)
2	0.74	0.72E-5	12	0.39E-4	24	(8)
3	0.59	0.13E-4	11	0.42E-4	30	(5)
4	0.74	0.14E-4	12	0.78E-4	49	(15)
5	0.59	0.25E-4	11	0.84E-4	60	(9)
6	0.42	0.57E-4	11	0.11E-3	91	(6)
7	0.74	0.24E-4	12	0.13E-3	80	(25)
8	0.59	0.43E-4	11	0.14E-3	100	(15)
9	0.42	0.94E-4	11	0.19E-3	152	(10)
10	0.74	0.38E-4	12	0.21E-3	129	(40)
11	0.59	0.68E-4	11	0.23E-3	161	(24)
12	0.42	0.15E-3	11	0.30E-3	242	(16)
13	0.74	0.48E-4	12	0.26E-3	161	(50)
14	0.59	0.85E-4	11	0.28E-3	199	(30)
15	0.42	0.19E-3	11	0.38E-3	305	(20)

Tables 5-1, 5-2 and 5-3 also give calculated apparent viscosities along with values extrapolated from rotating viscometer data. The latter appear in parentheses. Comparison shows that no correlation between the two exists, a result not entirely unexpected, since the measurements bear little relation to the downhole flow. On the other hand, calculated apparent viscosities correlated well with cuttings concentration, although not as well as did viscous stress. This correlation was possible because bottom-averaged shear rates did not vary appreciably from mud to mud at any given flow speed. This effect may be fortuitous.

Discussion 3:   Invert Emulsions versus "All Oil" Muds

Recently, Conoco's Jolliet project successfully drilled a number of deviated wells, ranging 30° to 60° from vertical, in the deepwater Green Canyon Block

Table 5-3
Bottom Averaged Fluid Properties @ 3.82 ft/sec

Mud	n	k lbf sec^n /in^2	Shear Rate 1/sec	Shear Stress (psi)	Apparent- Viscosity (cp)	
1	1.00	0.15E-6	18	0.27E-5	1	(1)
2	0.74	0.72E-5	16	0.49E-4	22	(8)
3	0.59	0.13E-4	15	0.50E-4	27	(5)
4	0.74	0.14E-4	17	0.98E-4	45	(15)
5	0.59	0.25E-4	15	0.10E-3	53	(9)
6	0.42	0.57E-4	15	0.13E-3	78	(6)
7	0.74	0.24E-4	16	0.16E-3	74	(25)
8	0.59	0.43E-4	15	0.17E-3	88	(15)
9	0.42	0.94E-4	15	0.21E-3	128	(10)
10	0.74	0.38E-4	16	0.26E-3	119	(40)
11	0.59	0.68E-4	15	0.27E-3	142	(24)
12	0.42	0.15E-3	15	0.34E-3	205	(16)
13	0.74	0.48E-4	17	0.33E-3	148	(50)
14	0.59	0.85E-4	15	0.33E-3	177	(30)
15	0.42	0.19E-3	15	0.43E-3	258	(20)

184 using a new "all oil" mud. Compared with wells previously drilled in the area with conventional invert emulsion fluids, the oil mud proved vastly superior with respect to cuttings transport and overall hole cleaning (Fraser, 1990a,b,c).

High levels of cleaning efficiency were maintained consistently throughout the drilling program. In this section we explain, using the fully predictive, eccentric annular flow model of Chapter 2, *why* the particular oil mud employed by Conoco performed well in comparison with the invert emulsion. The following discussion was first reported in Chin (1990c).

Given the success of the correlations developed in Discussions 1 and 2, it is natural to test our "stress hypothesis" under more realistic and difficult field conditions. Conoco's Green Canyon experience is ideal in this respect. Unlike the unweighted, bentonite-polymer, water-base muds used in the University of Tulsa experiments, the drilling fluids employed by Conoco were "invert emulsion" and "all oil" muds.

Again, Seeberger et al. (1989) have demonstrated how oil and water base muds having like rheograms, despite obvious textural differences, will clean holes with like efficiencies. This experimental observation implies that a knowledge of n and k alone suffices in characterizing the carrying capacity of water, oil or emulsion base drilling fluids. Thus, the use of a power law annular flow model as the basis for comparison for the two Conoco muds is completely warranted.

We assumed for simplicity a 2" radius drill pipe centered halfway down a 5" radius borehole. This eccentricity is consistent with the 30°- 60° inclinations reported by Conoco. The n and k values we required were calculated from Figure 2 of Fraser (1990b), using Fann dial readings at 13 and 50 rpm. For the invert emulsion, we obtained n = 0.55 and k = 0.0001 lbf sec^n/sq in; the values n = 0.21 and k = 0.00055 lbf sec^n/sq in were found for the "all oil" mud.

Our annular geometry is identical to that used in Discussion 1 and in Becker et al. (1989). It was chosen so that the shear stress results obtained for the Tulsa water-base muds (shown in Figures 5-1 to 5-3) can be directly compared with those found for the weighted invert emulsion and oil fluids considered here.

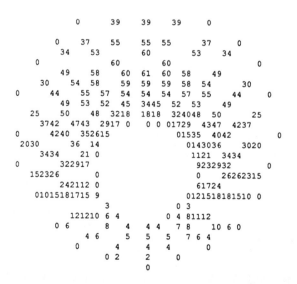

Figure 5-5a. Annular Velocity, Invert Emulsion.

```
        0       41    42    41     0
   0      40    47    47 47      40     0
      39    46        47        46    39
   0              47        47                  0
      46    47      47 47 47  47    46
   36    47  47      47 47 47  47  47      36
0      44    47  47  45  46  45  47  47    44    0
      46  46  45  42  3442  45  46    46
   33    46      44  3321  2121  334044  46      32
   4244  4543  3220 0    0 0 02032  4345  4341
0      4443  383019              01938  4244              0
  2837    40  18                0183540      3728
   3939    27 0                1527  4039
0      383523                13313538              0
   223133        0                0      33333122
      322919 0                102532
   0152325242214                0192226262315 0
            5                    0 5
      171815 9 6            0 6131718
      011      11    5    5 5  1011    1511 0
            6 9      7    7  7  11 9 6
         0        5    5  5      0
            0 3        3      0
                      0
```

Figure 5-5b. Annular Velocity, All Oil Mud.

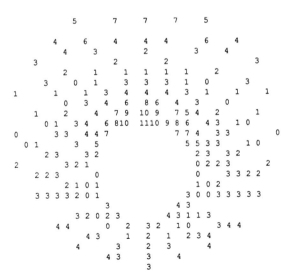

Figure 5-6a. Viscous Stress, Invert Emulsion.

For comparative purposes, the two runs described here were fixed at 500 gpm. To maintain this flow rate, the invert emulsion required an axial pressure gradient of 0.010 psi/ft; Conoco's all oil mud, by contrast, required 0.029 psi/ft. Figures 5-5a and 5-5b, for invert emulsion and all oil muds, give calculated results for axial velocity in terms of in/sec. Again, note how all no-slip conditions are identically satisfied.

Figures 5-6a and 5-6b display the absolute values of the *vertical* component of viscous shear stress; the leading significant digits are shown, corresponding to magnitudes that are typically $O(10^{-3})$ to $O(10^{-4})$ psi. This shear stress is obtained as the product of local apparent viscosity and shear rate, both of which vary throughout the cross-section. That is, the viscous stress is obtained *exactly* as "apparent viscosity (x,y) x dU(x,y)/dx".

Figure 5-5a shows that the invert emulsion yields maximum velocities near 61 in/sec on the high side of the annulus; the maximums on the low side, approximately 5 in/sec, are less than ten times this value. By comparison, the "all oil" results in Figure 5-5b demonstrate how a smaller n tends to redistribute velocity more uniformly; still, the contrast is high, being 47 in/sec to 7 in/sec. The difference between the low side maximum velocities of 5 and 7 in/sec is not significant, and certainly does *not* explain observed large differences in cleaning efficiency.

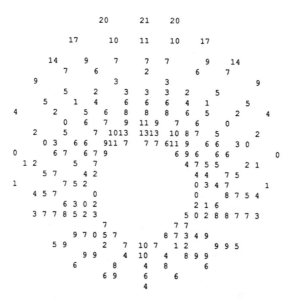

Figure 5-6b. Viscous Stress, All Oil Mud.

Our earlier results in Discussion 1 provided experimental evidence suggesting that mean viscous shear stress is the correct correlation parameter for hole cleaning efficiency. This is, importantly, again the case here. First note how Figure 5-6a gives a bottom radial stress distribution of "3-2-2-3-3" for the invert emulsion mud. In the case of Conoco's "all oil" mud, Figure 5-6b shows that these values significantly increase to "10-10-4-6-4".

We calculated mean shear stress values averaged over the lower half of the annulus. These values, for oil base and invert emulsion muds, respectively, were 0.00061 and 0.00027 psi. The corresponding ratio, a sizable 2.3, substantiates the positive claims made in Fraser (1990b). Calculated shear stress averages for the University of Tulsa experiments in no case exceeded 0.0004 psi.

Similarly averaged apparent viscosities also correlated well, leading to a large ratio of 2.2 (the "apparent viscosities" in Becker et al. (1989) did not correlate at all, because unmeaningful rotational viscometer readings were used). Bottom-averaged shear rates, for oil base and invert emulsion muds, were calculated as 12.7 and 9.6/sec, respectively; at least in this case, we have again justified the "6 rpm (or 10/sec) recommendation" offered by many drilling practitioners. In general, however, shear rates will vary widely; they can be substantial depending on the particular geometry and fluid.

The present results and the detailed findings of Discussion 1, together with the recommendations of Seeberger et al. (1989), strongly suggest that "bottom-averaged" viscous shear stress correlates well with cuttings carrying capacity. Thus, as before, a driller suspecting cleaning problems should first determine the current downhole stress level; then n, k and gpm should be altered to increase that stress. Once the danger is past, overall stress levels can be reduced to lower mud pump pressure requirements.

Discussion 4: Effect of Cuttings Bed Thickness

In vertical wells where chip movement is unimpeded, cuttings transport and hole cleaning efficiency vary directly as the product between "relative particle and annular velocity" and mud viscosity. For inclined wells, bed formation introduces a new physical source for clogging. Often, this means that rules of thumb developed for vertical holes are not entirely applicable to deviated wells. For example, Seeberger et al. (1989) pointed out that substantial increases in both yield point and annular velocity did not help in alleviating their hole problems. They suggested that high shear stresses at low shear rates would be desirable, and that stress could be a useful indicator of cleaning efficiency in deviated wells. We have given compelling evidence for this hypothesis.

Using the eccentric flow model of Chapter 2, we have demonstrated that "cuttings concentration" correlates *linearly* with mean shear stress; that is, *the viscous stress averaged over the lower half of the annulus,* for a wide range of oncoming flow speeds and well inclinations. Apparently, this empirical correlation holds for invert emulsions and oil base muds as well.

Having established that shear stress is an important parameter in bed formation, it is natural to ask consider whether cuttings bed growth itself helps or hinders further growth. That is, does bed buildup constitute a self-sustaining, destabilizing process? The classic "ball on top of the hill", for instance, continually falls once it is displaced from its equilibrium position. In contrast, the "ball in the valley" consistently returns to its origin, demonstrating "absolute stability".

If cuttings bed growth itself induces further growth, the cleaning process would be unstable in the foregoing sense. This instability would underline, in field applications, the importance of controlling downhole rheology so as to increase stress levels at the onset of impending danger. Field site flow simulation could play an important role in operations; that is, in determining existing stress levels with a view towards optimizing fluid rheology in order to

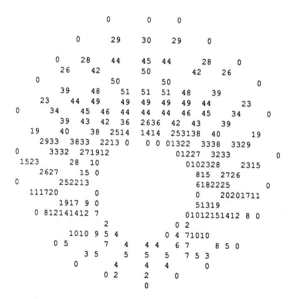

Figure 5-7a. Annular Velocity, "No Bed".

```
            0       0     0

       0      30    31    30     0

    0    29   45    46   45        29     0
      26    43        52         43    26
   0           51         51               0
       40    49    52   52   52   49    40
    23    45   50    50   51   50   50   45      23
  0    35    46   48   45   46   45   48   46    35      0
       40   44   43   37  2737  43   44    40
   19     41     39   2514 1414  253239  41        19
      3034  3834  2313 0   0  0  01323  3438  3430
  0     34     282012              01228    34             0
  1523    3228  10              0102332       2315
    2627     15  0              815   2726
  0       252212              6172225              0
  111719     0              0             191711
       191512 0              4121819
   0        6              0  6                0
       710121210   2          0  2  8111210  7
         6  5  1            1    6
         6     4  2     3  2     4  7  6
         5     4  3     3  3     4    5
         0  0  3  0     0  1  0  0     0
```

Figure 5-7b. Annular Velocity, "Small Bed".

```
            0       0     0

       0      30    31    30     0

    0    29   45    46   45        29     0
      26    43        52         43    26
   0           51         51               0
       40    49    52   52   52   49    40
    23    45   50    50   51   50   50   45      23
  0    35    46   48   45   46   45   48   46    35      0
       40   44   43   37  2737  43   44    40
   19     41     39   2514 1414  253239  41        19
      3034  3934  2313 0   0  0  01323  3439  3430
  0     34     282012              01228    34             0
  1523    3228  10              0102332       2315
    2727     15  0              815   2727
  0       252212              6182225              0
  111720     0              0             201711
       2015 9  0              5131920
   0        9  6              0  9                0
       711131311   2          0  2  131311  7
         8  4  2            1  2  7  8
         6     4  3  1     1  1  4  4  8  6
         0     0   0  1     0   0   2  0  4  0
```

Figure 5-7c. Annular Velocity, "Medium Bed".

increase them. In this section, calculations are described which suggest that instability is possible.

In the eccentric flow calculations that follow, we will assume a 2" radius nonrotating drill pipe, displaced 1.5" downward within a 5" radius borehole. This annular geometry is the same as the experimental set-up reported in Becker et al. (1989).

For purposes of evaluation, we arbitrarily selected "Mud No. 10" as described by the University of Tulsa team. It has a power law exponent of 0.736 and a consistency factor of 0.0000383 lbf sec^n/sq in. The total annular volume flow rate was fixed for all of our runs, corresponding to usual operating conditions. The average linear speed was held to 1.91 ft/sec or 22.9 in/sec. In the reported experiments, this speed yielded laminar flow at all inclination angles.

Four case studies were performed, the first containing no cuttings bed; then, assuming flat cuttings beds successively increasing in thickness. The level surfaces of the "small", "medium" and "large" beds were located at 0.4", 0.8" and 1.0", respectively, from the bottom of the annulus. Required pressure drops varied from 0.0054 to 0.0055 psi/ft. As indicated in Chapter 2, the highly visual output format directly overlays computed quantities on the cross-sectional geometry itself; thus, it facilitates physical interpretation and correlation with annular position.

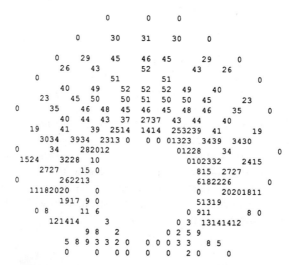

Figure 5-7d. Annular Velocity, "Large Bed".

Computed results for axial velocity in in/sec are shown in Figures 5-7a to 5-7d. All four velocity distributions satisfy the no-slip condition exactly; the crude plotter used, we note, does not always show 0s at solid boundaries because of character spacing issues. The "No Bed" flow given in Figure 5-7a demonstrates very clearly how velocity can vary rapidly about the annulus. For example, it has maximums of 51 and 5 in/sec above and below the pipe, a ten-fold difference. Figures 5-7b to 5-7d show that this factor increases, that is worsens, as the cuttings bed increases in thickness.

Figures 5-8a to 5-8d give computed results for the vertical component of the shear stress, that is "apparent viscosity (x,y) X strain rate dU/dx", where x increases downward. Results for the stress related to "dU/dy", not shown because of space limitations, behaved similarly. For clarity, only the absolute values are displayed; the actual values, which are separately available in tabulated form, vary from $O(10^{-4})$ to $O(10^{-3})$ psi.

Note how the bottom viscous stresses decrease in magnitude as the cuttings bed builds in thickness. This decrease, which is accompanied by decreases in through-put area, further compounds cuttings transport problems and decreases cleaning efficiency. Thus, hole clogging is a self-sustaining, destabilizing process. Unless the mud rheology itself is changed in the direction of increasing stress, differential sticking and stuck pipe are possible.

This decrease of viscous stress with increasing bed thickness is also supported experimentally. Quigley et al. (1990) measured "unexpected" decreases in fluid (as opposed to mechanical) friction in a carefully controlled flow loop where cuttings beds were allowed to grow. While concluding that "cuttings beds can reduce friction", the authors clearly do not recommend its application in the field, as it increases the possibility of differential sticking.

Numerical results such as those shown in Figures 5-8a to 5-8d provide a quantitative means for comparing cleaning capabilities between different muds at different flow rates. *"Should I use the `high tech' mud offered by Company A when the simpler drilling fluid of Company B, run at a different speed, will suffice?"* With numerical simulation, these and related questions are readily answered. The present results indicate that the smaller the throughput height, the smaller the viscous stresses will be. This is intuitively clear since narrow gaps impose limits upon the peak bottom velocity and hence the maximum stress. We caution that this result applies only to the present calculations and may not hold in general. The physical importance of cuttings beds indicates that they should be modelled in any serious well planning activity. This necessity also limits the potential of recently developed bipolar coordinate annular flow models. These handle circular eccentric annular geometries well, but they cannot be generalized to handle more difficult holes.

```
                55      57    55

            49      45    47    45    49

         39    39    28    29  28      39    39
            30    24          13      24    30
     27                12          12                  27
         17      9      4    4    4    9    17
      19      4    6    18   18   18   6    4        19
  12      8      9   19  31   31   31  19    9      8    12
          1   19   30   44   5844   30  19      1
      6    12    28   5169   7269  513628  12        6
      0 8   1924   436280   8380706243  2419   8 0
  1      1519   293250          555029  1915          1
    5 9      20  35              37352020        9 5
     1315      2018              1920   1513
  14        1514 8              4111415        14
   151517        1          1    15171515
       13 8 212              7 313
   23222018 9 210          21 3 21418202223
          21                2621
      1511 21518          2918 9 415
    2725        0   15   1815    6 0      212527
        2419        5    8    5  141924
        28          15   13   15      28
            2622        20      26
                        24
```

Figure 5-8a. Viscous Stress, "No Bed".

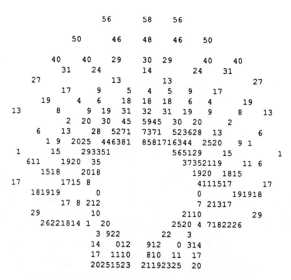

```
                56      58    56

            50      46    48    46    50

         40    40    29    30  29      40    40
            31    24          14      24    31
     27                13          13                  27
         17      9      5    4    5    9    17
      19      4    6    18   18   18   6    4        19
  13      8      9   19  31   32   31  19    9      8    13
          2   20   30   45   5945   30  20      2
      6    13    28   5271   7371  523628  13        6
      1 9   2025   446381   8581716344  2520   9 1
  1      15    293351          565129    15          1
    611      1920  35          37352119    11 6
     1518      2018              1920   1815
  17        1715 8              4111517        17
   181919        0              0      191918
       17 8 212              7 21317
   26221814 1 20          2520 4 7182226
          3 922          22    3
        14    012      912    0 314
        17    1110      810   11   17
        20251523   21192325  20
```

Figure 5-8b. Viscous Stress, "Small Bed".

```
            56        58    56

        50        46    48    46    50

      40    40    29    30  29      40      40
         31    24        14        24    31
    27              13        13                    27
       17      9      5    4    5    9    17
     19      4    6      18   18   18    6    4        19
  13      8      9   19   31   32   31   19    9      8      13
         2   20   30   45   5945  30   20      2
     6      13      28   5271  7371  523628  13          6
     1 8   2025   446381  8581716344  2520    8 1
  1      15      293351              565129      15              1
    510      1920   35              37352119      10 5
     1417      2018              1920  1714
  15        1714 8              4111417          15
   171818        1              1        181817
         17 8 212              7 31317
    27          410              21 4              27
    25211814 2  20              2520      8142125
            41414              1814 2 4
         19  14 3 0      4 5 0 8141019
         25   30   2013      6   20   26302225
```

Figure 5-8c. Viscous Stress, "Medium Bed".

```
            56        58    56

        50        46    48    46    50

      40    40    29    30  29      40      40
         31    24        14        24    31
    27              13        13                    27
       17      9      5    4    5    9    17
     19      4    6      18   18   18    6    4        19
  13      8      9   19   31   32   31   19    9      8      13
         2   20   30   45   5945  30   20      2
     6      12      28   5271  7371  523628  12          6
     1 8   2025   446381  8581716344  2520    8 1
  1      15      293351              565129      15              1
    510      1920   35              37352119      10 5
     1417      2018              1920  1714
  14        1714 8              4111417          14
   15171717        1              1        17171715
         13 8 212              7 313
    2423          210              21 3 2          2324
         211814      13              2413      8141821
            10 4      7              15 7 810
         2521152115 1 7      5 3 7 921  2125
         28      32   1711      5   17   2732      28
```

Figure 5-8d. Viscous Stress, "Large Bed".

Discussion 5:   Why 45⁰-60⁰ Inclinations are Worst

Various experimenters, for example, Becker et al. (1989) and Brown et al. (1989), have reported especially severe hole cleaning problems for deviated wells inclined approximately 45⁰ - 60⁰ from the vertical. The experimental data reported in the former paper, shown in our Figures 5-1, 5-2 and 5-3, indicate that cuttings concentration for a given flow speed peaks somewhere between ß = 30⁰ and 70⁰, where ß is measured from the vertical. These measurements appear to be reliable and repeatable; similar results have been reproduced at a number of test facilities.

One might ask why the expected worsening with increased inclination angle ß should *not* vary monotonically. Why should cuttings concentration at first increase, and then decrease? This relative maximum is easily understood. It follows from the fact that the *net* cuttings concentration C for a prescribed flow rate should vary smoothly as a function of ß; in particular, from a vertical value C_v at ß = 0⁰, to a horizontal value C_h at ß = 90⁰. That is, C must be weighted by resolving it into component contributions parallel and orthogonal to the well axis according to

$$C = C_h \sin ß + C_v \cos ß > 0 \tag{5-1}$$

To obtain relative maxima and minima, the usual rules of calculus require us to differentiate Equation 5-1 with respect to ß, and set the result to zero. That is, set

$$dC/dß = C_h \cos ß - C_v \sin ß = 0 \tag{5-2}$$

to obtain the critical angle and concentration

$$ß_{cr} = \arctan C_h/C_v \tag{5-3}$$

$$C_{max} = C_h/\sin ß_{cr} \tag{5-4}$$

That the critical concentration is a maximum is easily seen from the fact that

$$d^2C/dß^2 = - C_h \sin ß - C_v \cos ß = - C < 0 \tag{5-5}$$

is negative. Let us apply the data obtained of Becker et al. (1989). From Figure 5-3 previously, we estimate C_h = 25%; from Figure 5-1, we take C_v = 12%. Substitution in Equation 5-3 yields $ß_{cr}$ = 64⁰ and a corresponding C_{max} =

28%. The calculated 64⁰ agrees with general observation, while the 28% concentration is consistent with the high 45⁰ concentration results shown in Figure 5-2.

In general, once C_h and C_v are individually known from horizontal and vertical flow loop tests, it is possible using Equations 5-3 and 5-4 to determine the worst case inclination $ß_{cr}$ and its C_{max} for that particular mud and flow rate. In practice there may be some slight dependence of C_h on ß, since bed yield stresses may depend on gravity orientation and sedimentary packing. Note that ß is related to the α of Chapters 2, 3 and 4 by $α + ß = 90⁰$.

We emphasize that our "worst case" analysis applies only to unidirectional flows. When gravity segregation is important, the annular model developed in Chapter 4 may be more pertinent; the reverse flows possible for certain nondimensional channel parameters **Ch** only worsen as the the hole becomes more horizontal.

Discussion 6:   Additional Considerations

The empirical cuttings transport literature contains confusing observations and recommendations which, in light of the foregoing results and those of Chapter 3, can be easily resolved. References 12-21 provide a cross-section of recent experimental results and industry views, although our list is by no means exhaustive or comprehensive. We will address several questions commonly raised by drillers.

First and foremost is "Which parameters control cuttings transport efficiency?" In vertical wells, the drag or uplift force on small isolated chips can be obtained from lubrication theory via Stoke's or Oseen's low Reynolds number equations. This force is proportional to the product between local viscosity and the first power of relative velocity between chip and fluid. The so-called "Stokes product" correlates well in vertical holes.

In deviated and horizontal holes with eccentric annular geometries, cuttings beds invariably form on the low side. These beds consist of well-defined mechanical structures with nonzero yield stresses; to remove or erode them, viscous fluid stresses must be sufficiently strong to overcome their resilience. The stresses computed on a laminar basis are sufficient for practical purposes, because low side, low velocity flows are almost always laminar. In this sense, any turbulence in the high side flow is unimportant, since it plays no direct role in bed removal (the high side flow does convect debris that are uplifted by rotation). This observation is reiterated by Fraser (1990c). In his paper, Fraser correctly points out that too much significance is often attached to velocity criteria and fluid turbulence in deviated wells.

A second common question concerns the role of drillpipe rotation. With rotation, centrifugal effects throw cuttings circumferentially upwards where they are convected uphole by the high side flow; then they fall downwards. In the first part of this cycle, the cuttings are subject to drag forces not unlike those found in vertical wells. Here turbulence can be important, determining the amount of axial throw traversed before the cuttings are redeposited into the bed. Order of magnitude estimates comparing rotational to axial effects can be obtained using the formulas in Chapter 3.

Other effects of rotation are subtly tied to the rheology of the carrier fluid. Conflicting observations and recommendations are often made regarding drillpipe rotation for concentric annuli. To resolve them, we need to reiterate some theoretical results of Chapter 3. There we demonstrated that axial and circumferential speeds completely decouple for laminar Newtonian flows despite the nonlinearity of the Navier-Stokes equations. This is so because the convective terms exactly vanish, allowing us to "naively" superpose the two orthogonal velocity fields.

This fact was, apparently, first deduced by Savins and Wallick (1966), who noted that no coupling between the discharge rate, axial pressure gradient, relative rotation, and torque could be found through the viscosity coefficient for Newtonian flows. This author is indebted to J. Savins for directing him to the earlier literature.

The decoupling implies that experimental findings obtained using Newtonian drilling fluids (primarily water and air) cannot be extrapolated to more general power law or Bingham plastic rheologies. Likewise, rules of thumb deduced using real drilling muds will not be consistent with those found for water. Newtonian and "real" muds behave differently in the presence of pipe rotation. In a Newtonian fluid, rotation will not affect the axial flow, although centrifugal "throwing" is still important.

In an initially steady non-Newtonian flow where the mudpump is operating at constant pressure, a momentary increase in rpm leads to a temporary surge in flow rate and thus improved hole cleaning. However, once the pump readjusts itself to the prescribed gpm, this advantage is lost; unless, obviously, the discharge rate itself is reset upwards or the rheology is improved by using a mud additive.

The decoupling discussed above applies to Newtonian flows in concentric annuli only. The coupling between axial and circumferential velocities reappears - even for Newtonian flows - when the rotating motion occurs in an eccentric annulus. This is so because the nonlinear convective terms will not identically vanish. This isolated singularity suggest that concentric flow loop tests using Newtonian fluids provide little benefit in terms of field usefulness. Their results are, moreover, subject to misinterpretation.

And the role of fluid rheology? We have demonstrated how bottom-averaged shear stress can be used as a meaningful correlation parameter for cuttings transport in eccentric deviated holes. This mean viscous stress can be computed using the method developed in Chapter 2. The arguments given in Discussions 1, 2 and 3 are sound on physical grounds; in cuttings transport, rheology is a significant player by way of its effect on fluid stress.

We emphasize that we have *not* modelled the dynamics of single chips, or collective ensembles of cuttings. Such models are common in civil and chemical engineering, where the dynamics of well defined individual bodies are sought. In drilling, this is hardly the case. Nor are such analyses recommended: for field applications, it is only necessary to use stress as a correlation parameter. Finally, a comment on the statement that increased fluid density improves hole cleaning. This is undeniably the rule, since higher densities increase buoyancy effects; it applies to all flows, whether or not they are annular or deviated.

EVALUATION OF SPOTTING FLUIDS FOR STUCK PIPE

Stuck pipe due to differential pressure between the mud column and the formation often results in costly time delays. The mechanics governing differential sticking are well known (Outmans, 1958). In the past, diesel oil, mineral oil, and mixtures of these with surfactants, clays and asphalts were usually spotted to facilitate the release of the drill string. However, the use of these conventional spotting fluids is now stringently controlled by government regulation; environmentally safe alternatives must be found.

Recently, Halliday and Clapper (1989) described the development of a successful, non-toxic water-base system. Such systems are becoming increasingly important. Their new spotting fluid, identified using simple laboratory screening procedures, was used to free a thousand feet of stuck pipe in a 39⁰ hole from a sand section in the Gulf of Mexico.

Since water base spotting fluids, being relatively new, have seldom been studied in the literature, it is natural to ask whether or not they really work; and, if so, how. This section calculates on an eccentric flow basis three important mechanical properties; namely, the apparent viscosity, shear stress and shear rate of the drilling mud, with and without the spot additive. Then we provide a complete physical explanation for the reported success. The spotting fluid essentially works by mechanically reducing overall apparent viscosity; this enables the resultant fluid to better perform its chemical functions. The results of this section were first reported in Chin (1991).

The eccentric borehole annular flow model of Chapter 2 was used to answer several questions related to the effectiveness of spotting fluids in freeing stuck pipe. Which mechanical properties are relevant to spotted fluids? What should their orders of magnitude be? Note that it suffices to explain how the water base

Table 5-4
Fluid Properties, Original Mud

Pressure Gradient (psi/ft)	Flow Rate (gpm)	Apparent-Viscosity (lbf sec/in^2)	Shear Rate (sec^{-1})	Viscous Stress (psi)
0.0010	69	0.000036	0.4	0.000011
0.0020	185	0.000027	1.2	0.000022
0.0030	329	0.000022	2.1	0.000033
0.0035	410	0.000021	2.6	0.000038
0.0040	497	0.000020	3.2	0.000044
0.0050	683	0.000018	4.3	0.000055
0.0060	886	0.000017	5.6	0.000066
0.0070	1105	0.000016	7.0	0.000077

Table 5-5
Fluid Properties, Spotted Mud

Pressure Gradient (psi/ft)	Flow Rate (gpm)	Apparent-Viscosity (lbf sec/in^2)	Shear Rate (sec^{-1})	Viscous Stress (psi)
0.0010	140	0.000014	1.0	0.000011
0.0020	344	0.000012	2.4	0.000022
0.0023	412	0.000011	2.8	0.000025
0.0030	582	0.000010	4.0	0.000033
0.0035	711	0.000010	4.9	0.000039
0.0040	846	0.000010	5.8	0.000044
0.0050	1130	0.000009	7.8	0.000055

spotting fluid behaves, insofar as mechanical fluid properties are concerned, on a single-phase, miscible flow basis. Conventional capillary pressure and multiphase considerations for "oil on aqueous filter cake" effects do not apply here, since we are dealing with "water on water" flows.

We performed our calculations for a 7.75" diameter drill collar located eccentrically within a 12.5" diameter borehole. This corresponds to the bottomhole assembly reported by the authors. A small bottom annular clearance of 0.25" was selected for evaluation purposes. This almost closed gap is consistent with the impending stuck pipe conditions characteristic of typical deviated holes.

The authors' Table 11 gives Fann 600 and 300 rpm dial readings for the water base mud used, before and after spot addition; both fluids, incidentally, were equal in density. In the former case, these values were 46 and 28; in the latter, 41 and 24. These properties were measured at 120° F. The calculated n and k power law coefficients are, respectively, 0.70 and 0.000025 lbf sec^n/sq in for the original mud; for the spotted mud, we obtained 0.77 and 0.0000137 lbf sec^n/sq in.

Halliday and Clapper reported that attempts to free the pipe by jarring down, with the original drilling fluid in place, were unsuccessful. At that point, the decision to spot the experimental non-oil fluid was made. Since jarring operations are more impulsive, rather than constant pressure drop processes, we calculated our flow properties for a wide range of applied pressure gradients. Note that the unsteady, convective term in the governing momentum equation has the same physical dimensions as pressure gradient. It was in this approximate engineering sense that our exact simulator was used. The highest pressure gradients shown below correspond to instantaneous volume flow rates near 1100 gpm. Computed results for several parameters averaged over the lower half of the annulus are shown in Tables 5-4 and 5-5.

We emphasize that calculated averages are sensitive to annular geometry; thus, the results shown in Tables 5-4 and 5-5 may not apply to other borehole configurations. In general, any required numerical quantities should be recomputed with the exact downhole geometry.

The results for averaged shear stress are "almost" Newtonian, in the sense that stress increases linearly with applied pressure gradient. This unexpected outcome is not generally true of non-Newtonian flows. Both treated and untreated muds, in fact, show exactly the same shear stress values. However, shear rate and volume flow rate results for the two muds vary differently, and certainly nonlinearly with pressure gradient. The most interesting results, those concerned with spotting properties, are related to apparent viscosity.

The foregoing calculations importantly show how the apparent viscosity for the spotted mud, which varies spatially over the annular cross-section, has a nearly constant "bottom average" near 0.000010 lbf sec/in^2 over the entire range of flow rates. This value is approximately 69 cp, far in excess of the viscosities inferred from rotational viscometer readings, but still *two to three times less* than those of the original untreated mud. The importance of "low viscosity" in spotting fluids is emphasized in several mud company publications brought to this author's attention. Whether the apparent viscosity is high or low, of course, *cannot* be determined independently of the hole geometry and the prescribed pressure gradient.

The apparent viscosity is relevant because it is related to the lubricity factor conventionally used to evaluate spotting fluids. It is importantly calculated on an eccentric flow basis, rather than determined from an (unrelated) rotational viscometer measurement. As in cuttings transport, viscometer measurements are only valid to the extent that they provide accurate information for determining n and k over a limited range of shear rates.

That the treated fluid exhibits much lower viscosities over a range of applied pressures is consistent with its ability to penetrate the pipe and mudcake interface. This lubricates and separates the contact surfaces over a several hour period; thus it enables the spotting to perform its chemical functions efficiently, thereby freeing the stuck drill string. The effectiveness of any spotting fluid, of course, must be determined on a case by case basis.

While computed averages for apparent viscosity are almost constant over a range of pressure gradients, we emphasize that exact cross-sectional values for each flow property can be quite variable. For example, consider the annular flow for the spotted mud under a pressure gradient of 0.002 psi/ft, with a corresponding flow rate of 344 gpm. The velocity solutions in in/sec, using the highly visual output format discussed in Chapter 2, are shown in Figure 5-9; note, again, how no-slip conditions are rigorously enforced at all solid surfaces.

Figure 5-10 gives results for *exact apparent viscosity*, which varies with spatial position, again plotted over the highly eccentric geometry itself. Although the crude text plotter used does not furnish sufficient visual resolution at the bottom of the annular gap, reference to tabulated solutions indicates pipe surface values of "13", increasing to "29" at the midsection, finally decreasing to 13×10^{-6} lbf sec/in^2 at the borehole wall. The flatness of the cuttings bed, or the extent to which it modifies annular bottom geometry, will also be an important factor as far as lubricity is concerned. Any field oriented hydraulics simulation should also account for such bed effects.

We had demonstrated earlier that flow modeling can be used to correlate laboratory and field cuttings transport efficiency data against actual (computed)

```
                    0        0    0
              0     21       22   21     0
                             35
        0     20        34   41   34    20     0
                   32   40   43   40    32
              17   37   42   42   42    37    17
        0    2833 3938 38   39   38    39   3328      0
           14    34    35 32 33   32  35    34    14
           23    33    30 24 13   24  30    33    23
        0    27   30   22   0    0   0  22   30    27      0
        11    27  1810 0              0   18   27    11
       1720  2420   0                      0    24  2017
        20    14                               14    20
        0    1914 0                            01419      7 0
        11   10                                10    11
         13     0                              0    13
       0 4  11 6                              611      4 0
         6 7   3                                3    7 6
           7 4                                  4 7
         0 2   3 0                            0 3     2 0
           3 3 3                                3 3 3
              2                                 1 2
           0 0 1 1                            1 1 0 0
                0                                0 0
              0 0 0                            0 0 0
                    0 0        0    0 0
```

Figure 5-9. Annular Velocity.

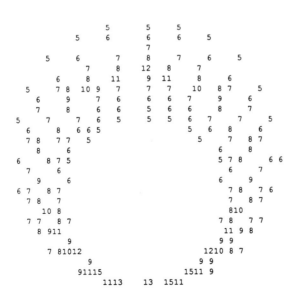

Figure 5-10. Apparent Viscosity.

downhole flow properties. Bottom-averaged viscous shear stress importantly emerged as the physically significant correlation parameter. The present section indicates that annular flow modeling can also be used to evaluate the effectivenes of spotting fluids in freeing stuck pipe. Here, the important correlation parameter is average apparent viscosity, a fact most mechanical engineers might have anticipated. This quantity is directly related to the lubricity factor usually obtained in laboratory measurements.

CEMENTING APPLICATIONS

The modeling of annular borehole flows for drilling applications is similar to that for cementing. Aside from obvious differences associated with "touch and feel" contrasts between drilling muds and cement slurries, little in the way of analysis changes. What differs, however, lies in the way computed quantities are used. In drilling problems, "mean viscous shear stress" and average velocity control cuttings transport efficiency, depending on the deviation of the well. On the other hand, a low value of "mean apparent viscosity" appears to determine the effectiveness of a spotted mud for use in releasing stuck pipe.

References 26-39 provide a representative cross-section of the modern cementing literature. The primary operational concern is effective mud displacement and removal. The industry presently emphasizes the importance of good rheology and high velocity, but these qualities alone are not sufficient. In order to produce good displacements, the stability of the cement velocity profile with respect to disturbances induced by the upstream mud should be addressed.

The cement velocity profile must be hydrodynamically stable and robust. Unstable velocity distributions may break down rapidly into viscous fingers and channel prematurely. An analogous problem in reservoir engineering is found in waterflooding: the displacement front may disperse into tiny fingers that propagate into the downstream flow when adverse mobility ratios are encountered. There the problem is solved by using flow additives whose attributes are determined by detailed mathematical modeling.

The classic monograph of Lin (1967) explains how the stability characteristics of any particular flow can be obtained as solutions to the so-called Rayleigh or Orr-Sommerfeld equations. Good velocity profiles in this sense can be ascertained by coupling the work of Chapter 2 which generates velocity profiles, to stability models that evaluate their ability to withstand disturbances. Stability analyses are routinely used in aeronautical and chemical engineering; for example, in the study of turbulent transition on wing surfaces and in ducts. They are also important in different secondary recovery aspects of reservoir engineering. More research should be directed towards this area.

In the remainder of this section, typical velocity profiles are generated for comparative purposes only using the models of Chapters 2 and 3. These are not evaluated with respect to hydrodynamic stability.

Example 1:   Eccentric Nonrotating Flow, Baseline Concentric Case

---

We first establish a simple concentric solution as the basis for further annular flow comparison. We will assume for the casing outer diameter a "pipe radius" of 3.0 in and a borehole radius of 4.5 in. We will evaluate the behaviour of two cement slurries. The first is an API Class H slurry with power law coefficients $n = 0.30$ and $k = 0.001354$ lbf sec^n/in^2, while the second is a Class C slurry having $n = 0.43$ and $k = 0.0002083$ lbf sec^n/in^2. We shall refer to these as our "Class H" and "Class C" flows.

For convenience only, we will fix in our comparisons throughout an approximate number for the pressure gradient, say $dP/dz = -0.1$ psi/ft. Thus some predicted gpms may be excessive from an engineering standpoint. The simulations described in Examples 1-4 below are run "as is" in the comparative sense of Chapter 1, and no attempt has been made to "fine tune" or calibrate the variable mesh to any known solution. The computed velocity distribution in in/sec for the Class H slurry is shown in Figure 5-11a. The corresponding volume flow rate is 8.932 gpm. The stress formed by the product "apparent viscosity (x,y) x $dU(x,y)/dx$" appears in Figure 5-11b. Typically, these stresses might be 0.5×10^{-3} psi in magnitude. A run summary is given in Table 5-6.

Next, we will rerun the simulation for the Class C slurry. The volume flow rate obtained in this case is 849.1 gal/min. The computed velocity and stress profiles are shown in Figure 5-11c (where the "12" indicates 120 in/sec) and Figure 5-11d. Run summaries for averaged quantities are given in Table 5-7.

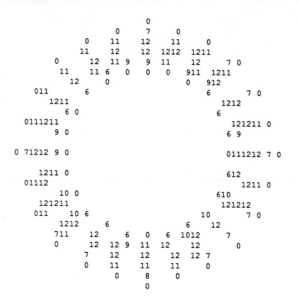

Figure 5-11a. Annular Velocity, Class H Slurry.

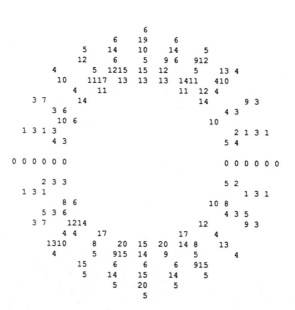

Figure 5-11b. Viscous Stress, Class H Slurry.

```
                          0
                 0        6        0
          0     10       12      10       0
         10     12       12    1212    1210
    0       12   11 9    9    11      12        6 0
   10      11 5    0    0     0      911  1210
    12      0                        0    912
  010        5                         5          6 0
   1211                                1212
     5 0                                 5
0101211                                  121210 0
   9 0                                   5 9

0 61212 9 0                               0111212 6 0

   1211 0                                512
01012                                     1210 0
     9 0                                5 9
  121211                                121212
   010    9 5                          9        6 0
    1212       5                    5     12
     610     12      5    0    5    912      6
      0      12   12 9   11   12     12        0
       6     12      12      12   12 6
        0    10      10      10       0
              0       7       0
                      0
```

Figure 5-11c. Annular Velocity, Class C Slurry.

```
                          19
                 19       24      19
         17      17       12      17      17
         15       6        5   10 6   1015
   14       6   1520   21   15       6      1714
    12      1322  23   24   23    1813     412
     3      21                   21   15 3
  9 9      18                       18       12 9
    3 8                                5 3
    1312                             13
5 4 1 4                                2 1 4 5
    5 6                                6 5

0 0 0 0 0 0                            0 0 0 0 0 0

    2 4 6                              6 2
5 4 1                                    1 4 5
     1012                             1310
    6 3 8                              5 3 6
  9 9   1518                        15        12 9
     4 4      22                  22      4
   1712      9   25   27   25   18 9     17
    14       5  1121   18   11     5      14
        20       6        6    6  1020
        17      17       19      17   17
                18       25      18
                         20
```

Figure 5-11d. Viscous Stress, Class C Slurry.

Table 5-6
Example 1: Summary, Average Quantities (Class H Slurry)

TABULATION OF CALCULATED AVERAGE QUANTITIES:
Area weighted means of absolute values taken over
BOTTOM HALF of annular cross-section ...
O Average annular velocity = .8955E+00 in/sec
O Average apparent viscosity = .2378E-02 lbf sec/sq in
O Average stress, AppVis x dU/dx, = .8820E-03 psi
O Average stress, AppVis x dU/dy, = .7530E-03 psi
O Average dissipation = .3112E-02 lbf/(sec sq in)
O Average shear rate dU/dx = .1500E+01 1/sec
O Average shear rate dU/dy = .1278E+01 1/sec
O Average Stokes product = .2908E-02 lbf/in

TABULATION OF CALCULATED AVERAGE QUANTITIES, II:
Area weighted means of absolute values taken over
ENTIRE annular (x,y) cross-section ...
O Average annular velocity = .8937E+00 in/sec
O Average apparent viscosity = .2397E-02 lbf sec/sq in
O Average stress, AppVis x dU/dx, = .8030E-03 psi
O Average stress, AppVis x dU/dy, = .7964E-03 psi
O Average dissipation = .3076E-02 lbf/(sec sq in)
O Average shear rate dU/dx = .1362E+01 1/sec
O Average shear rate dU/dy = .1349E+01 1/sec
O Average Stokes product = .2925E-02 lbf/in

Table 5-7
Example 1: Summary, Average Quantities (Class C Slurry)

TABULATION OF CALCULATED AVERAGE QUANTITIES:
Area weighted means of absolute values taken over
BOTTOM HALF of annular cross-section ...
O Average annular velocity = .8511E+02 in/sec
O Average apparent viscosity = .1816E-04 lbf sec/sq in
O Average stress, AppVis x dU/dx, = .1190E-02 psi
O Average stress, AppVis x dU/dy, = .1015E-02 psi
O Average dissipation = .4555E+00 lbf/(sec sq in)
O Average shear rate dU/dx = .1436E+03 1/sec
O Average shear rate dU/dy = .1223E+03 1/sec
O Average Stokes product = .2027E-02 lbf/in

TABULATION OF CALCULATED AVERAGE QUANTITIES, II:
Area weighted means of absolute values taken over
ENTIRE annular (x,y) cross-section ...
O Average annular velocity = .8497E+02 in/sec
O Average apparent viscosity = .1826E-04 lbf sec/sq in
O Average stress, AppVis x dU/dx, = .1082E-02 psi
O Average stress, AppVis x dU/dy, = .1073E-02 psi
O Average dissipation = .4502E+00 lbf/(sec sq in)
O Average shear rate dU/dx = .1304E+03 1/sec
O Average shear rate dU/dy = .1291E+03 1/sec
O Average Stokes product = .2036E-02 lbf/in

Example 2: Eccentric Nonrotating Flow,
Eccentric Circular Case

The same pipe and hole sizes used in Example 1 are assumed here, but the casing is displaced downward, resting off the bottom by 0.5 in. Both the casing and the borehole are perfect circles. For the Class H slurry, a flow rate of 48.4 gpm is obtained, exceeding the 8.93 gpm of Example 1. Self-explanatory computed results are given for comparative purposes. In Figure 5-12a the "9" indicates 9 in/sec, while in Figure 5-12b the stresses are typically 0.001 psi. Averaged results appear in Table 5-8.

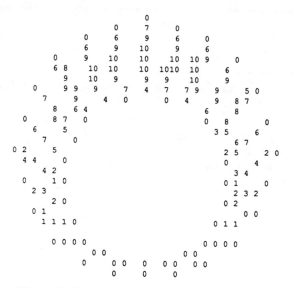

Figure 5-12a. Annular Velocity, Class H Slurry.

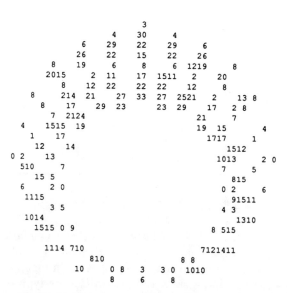

Figure 5-12b. Viscous Stress, Class H Slurry.

Table 5-8
Example 2: Summary, Average Quantities (Class H Slurry)

TABULATION OF CALCULATED AVERAGE QUANTITIES:
Area weighted means of absolute values taken over
BOTTOM HALF of annular cross-section ...
O Average annular velocity = .5958E+00 in/sec
O Average apparent viscosity = .2206E-02 lbf sec/sq in
O Average stress, AppVis x dU/dx, = .8796E-03 psi
O Average stress, AppVis x dU/dy, = .1014E-02 psi
O Average dissipation = .4405E-02 lbf/(sec sq in)
O Average shear rate dU/dx = .1097E+01 1/sec
O Average shear rate dU/dy = .1796E+01 1/sec
O Average Stokes product = .5406E-03 lbf/in

TABULATION OF CALCULATED AVERAGE QUANTITIES, II:
Area weighted means of absolute values taken over
ENTIRE annular (x,y) cross-section ...
O Average annular velocity = .3470E+01 in/sec
O Average apparent viscosity = .1302E-02 lbf sec/sq in
O Average stress, AppVis x dU/dx, = .1171E-02 psi
O Average stress, AppVis x dU/dy, = .1201E-02 psi
O Average dissipation = .1560E-01 lbf/(sec sq in)
O Average shear rate dU/dx = .4129E+01 1/sec
O Average shear rate dU/dy = .4283E+01 1/sec
O Average Stokes product = .2691E-02 lbf/in

The calculations for this circular eccentric casing and hole are now repeated for the Class C slurry. Here, the volume flow rate is 2,706 gal/min. Computed results for the velocity distribution are shown in Figure 5-12c, where "60" indicates 600 in/sec; averaged results are recorded in Table 5-9.

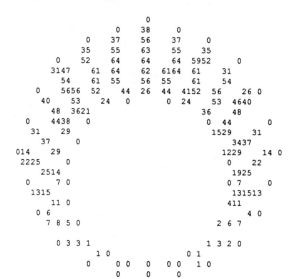

Figure 5-12c. Annular Velocity, Class C Slurry.

Table 5-9
Example 2: Summary, Average Quantities (Class C Slurry)

TABULATION OF CALCULATED AVERAGE QUANTITIES:
Area weighted means of absolute values taken over
BOTTOM HALF of annular cross-section ...
O Average annular velocity = .3340E+02 in/sec
O Average apparent viscosity = .2407E-04 lbf sec/sq in
O Average stress, AppVis x dU/dx, = .8455E-03 psi
O Average stress, AppVis x dU/dy, = .9885E-03 psi
O Average dissipation = .2660E+00 lbf/(sec sq in)
O Average shear rate dU/dx = .6535E+02 1/sec
O Average shear rate dU/dy = .9999E+02 1/sec
O Average Stokes product = .5218E-03 lbf/in

TABULATION OF CALCULATED AVERAGE QUANTITIES, II:
Area weighted means of absolute values taken over
ENTIRE annular (x,y) cross-section ...
O Average annular velocity = .1930E+03 in/sec
O Average apparent viscosity = .1542E-04 lbf sec/sq in
O Average stress, AppVis x dU/dx, = .1366E-02 psi
O Average stress, AppVis x dU/dy, = .1369E-02 psi
O Average dissipation = .1176E+01 lbf/(sec sq in)
O Average shear rate dU/dx = .2361E+03 1/sec
O Average shear rate dU/dy = .2333E+03 1/sec
O Average Stokes product = .2112E-02 lbf/in

Example 3:   Eccentric Nonrotating Flow, A Severe Washout

Here the eccentric geometry of Example 2 is modified by including a severe washout at the top of the hole. The washout is nonsymmetrical; it may, for example, have resulted from the rotating action of the drillbit against an unconsolidated sand. We could just as easily have modeled a keyseat indentation or any other wall deformation, of course. Average properties for the Class H and Class C flows are listed in Tables 5-10 and 5-11 respectively.

Table 5-10
Example 3: Summary, Average Quantities (Class H Slurry)

TABULATION OF CALCULATED AVERAGE QUANTITIES:
Area weighted means of absolute values taken over
BOTTOM HALF of annular cross-section ...
O Average annular velocity = .8670E+00 in/sec
O Average apparent viscosity = .1982E-02 lbf sec/sq in
O Average stress, AppVis x dU/dx, = .9911E-03 psi
O Average stress, AppVis x dU/dy, = .1123E-02 psi
O Average dissipation = .7581E-02 lbf/(sec sq in)
O Average shear rate dU/dx = .1636E+01 1/sec
O Average shear rate dU/dy = .2620E+01 1/sec
O Average Stokes product = .5448E-03 lbf/in

TABULATION OF CALCULATED AVERAGE QUANTITIES, II:
Area weighted means of absolute values taken over
ENTIRE annular (x,y) cross-section ...
O Average annular velocity = .1338E+02 in/sec
O Average apparent viscosity = .9957E-03 lbf sec/sq in
O Average stress, AppVis x dU/dx, = .1683E-02 psi
O Average stress, AppVis x dU/dy, = .1720E-02 psi
O Average dissipation = .8241E-01 lbf/(sec sq in)
O Average shear rate dU/dx = .1402E+02 1/sec
O Average shear rate dU/dy = .1285E+02 1/sec
O Average Stokes product = .2727E-02 lbf/in

Table 5-11

Example 3: Summary, Average Quantities (Class C Slurry)

TABULATION OF CALCULATED AVERAGE QUANTITIES:
Area weighted means of absolute values taken over
BOTTOM HALF of annular cross-section ...
O Average annular velocity = .3685E+02 in/sec
O Average apparent viscosity = .2351E-04 lbf sec/sq in
O Average stress, AppVis x dU/dx, = .8819E-03 psi
O Average stress, AppVis x dU/dy, = .1025E-02 psi
O Average dissipation = .3129E+00 lbf/(sec sq in)
O Average shear rate dU/dx = .7239E+02 1/sec
O Average shear rate dU/dy = .1105E+03 1/sec
O Average Stokes product = .5244E-03 lbf/in

TABULATION OF CALCULATED AVERAGE QUANTITIES, II:
Area weighted means of absolute values taken over
ENTIRE annular (x,y) cross-section ...
O Average annular velocity = .4319E+03 in/sec
O Average apparent viscosity = .1338E-04 lbf sec/sq in
O Average stress, AppVis x dU/dx, = .1718E-02 psi
O Average stress, AppVis x dU/dy, = .1781E-02 psi
O Average dissipation = .3035E+01 lbf/(sec sq in)
O Average shear rate dU/dx = .4546E+03 1/sec
O Average shear rate dU/dy = .4245E+03 1/sec
O Average Stokes product = .2495E-02 lbf/in

The flow rate for the Class H slurry is 274.9 gpm, exceeding the 48.4 gpm of Example 2. The corresponding velocity results are displayed in Figure 5-13a, where "50" indicates 50 in/sec. The simulations for the Class C slurry gave a high volume flow rate of 8,819 gpm; detailed velocity distributions are shown in Figure 5-13b where "20" means 2000 in/sec.

```
                              0
                0
                            Severe Washout

                0    58    61    0
                    39    66
            0              67    44
          27    53    67    67
          41    58    67    67    55
      0   46    59    65    65    58      0
      1828    47    54    61    60    58    21
        33    46  4837    42    53    51    37
    0   3434    37    21    24    24  3344    39      14 0
      20      31    15    0        0   19    35   2622
        24  2011                      23    27
    0   2318    0                      0  2326        0
      13    13                          815    14
        15      0                        0  1417
  0 5     11                              512      5 0
    8 9      0                            0      8
      8 4                                 6 9
    0     2 0                            0 2        0
      4 4                                 4 4 4
        3 0                               1 3
    0 1                                         1 0
      2 2 1 0                            0 1 2

        0 0 0 0                      0 0 0 0
            0 0                    0 0
          0 0    0 0      0    0 0    0 0
              0        0        0
                  0
```

Figure 5-13a. Annular Velocity, Class H Slurry.

Example 4: Eccentric Nonrotating Flow, Casing with Centralizers

----

Centralizers are typically used to prevent excessively low velocity cement zones from forming; the annular flow is concentric, more or less, and eccentricity is avoided at all costs. In this final example, we reconsider the concentric annular borehole flow of Example 1; however, we will introduce four centralizers. The geometry is essentially the same for drill collars with stabilizers. Again, the pipe and borehole radii are, respectively, 3.0 in and 4.5 in. For clarity, the screen displays for both the annular geometry and its corresponding computational mesh are duplicated in Figures 5-14a and 5-14b. Note how our centralizers are modelled by indenting the borehole contour inward. Alternatively, we could have deformed the pipe contour outward.

For the Class H slurry, the computed volume flow rate for this configuration is 8.30 gpm, which is slightly less than the value obtained without centralizers. This is physically consistent with the blockage effects introduced by the centralizers. Finally for the Class C slurry, the volume flow rate is 645.8 gpm. Velocity results are shown in Figures 5-14c ("15" means 1.5 in/sec) and 5-14d ("12" is 120 in/sec), and flowfield averages are given in Tables 5-12 and 5-13.

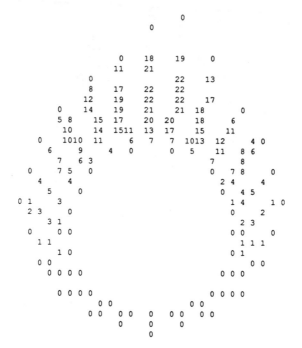

Figure 5-13b. Annular Velocity, Class C Slurry.

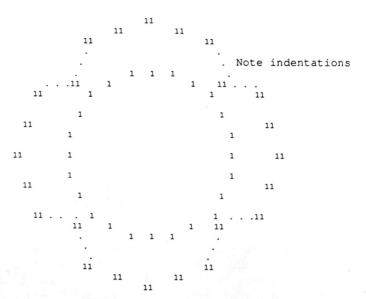

Figure 5-14a. Concentric Casing and Hole with Centralizers.

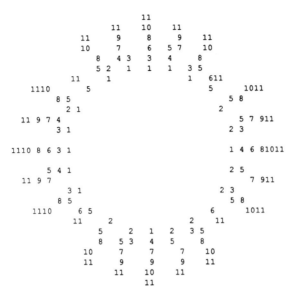

Figure 5-14b. Centralizer Fitted Mesh System.

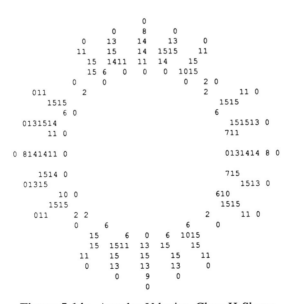

Figure 5-14c. Annular Velocity, Class H Slurry.

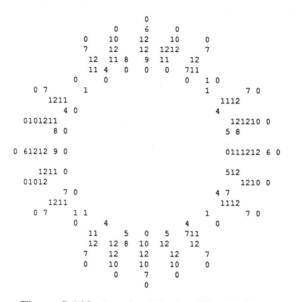

Figure 5-14d. Annular Velocity, Class C Slurry.

Taking n and k values typical of commonly used cement slurries, we have shown how velocity profiles for very general annular geometries can be computed in a stable and efficient manner. Also, the correct qualitative trends are captured; for example, increasing flow rate with eccentricity, decreasing volume flow with centralizer blockage. Simulation again allows us to answer "what if" questions quickly and inexpensively. What is the effect of a washout on bottom velocity? What effects will centralizer width have on velocity peaks? How does rheology interact with annular geometry at specific flow rates or pressure drops?

Again the primary operational concern in cementing is effective mud displacement and removal. Cement is known to channel through drilling mud, leaving mud behind. This may require expensive remedial work. But channeling is a hydrodynamic stability phenomenon that has been amply studied in the engineering literature. It is possible, as is routinely done in reservoir engineering as well as outside industries, to evaluate the ability of any particular velocity profile to remain stable to flow disturbances. This robustness - which can be tailored by changing rheological properties or geometry - should be properly exploited to control undesirable fingering or channeling.

Table 5-12
Example 4: Summary, Average Quantities (Class H Slurry)

TABULATION OF CALCULATED AVERAGE QUANTITIES:
Area weighted means of absolute values taken over
BOTTOM HALF of annular cross-section ...
O Average annular velocity = .9092E+00 in/sec
O Average apparent viscosity = .1691E-02 lbf sec/sq in
O Average stress, AppVis x dU/dx, = .1063E-02 psi
O Average stress, AppVis x dU/dy, = .9636E-03 psi
O Average dissipation = .4090E-02 lbf/(sec sq in)
O Average shear rate dU/dx = .1682E+01 1/sec
O Average shear rate dU/dy = .1447E+01 1/sec
O Average Stokes product = .1840E-02 lbf/in

TABULATION OF CALCULATED AVERAGE QUANTITIES, II:
Area weighted means of absolute values taken over
ENTIRE annular (x,y) cross-section ...
O Average annular velocity = .9185E+00 in/sec
O Average apparent viscosity = .1674E-02 lbf sec/sq in
O Average stress, AppVis x dU/dx, = .9290E-03 psi
O Average stress, AppVis x dU/dy, = .9398E-03 psi
O Average dissipation = .3970E-02 lbf/(sec sq in)
O Average shear rate dU/dx = .1519E+01 1/sec
O Average shear rate dU/dy = .1514E+01 1/sec
O Average Stokes product = .1852E-02 lbf/in

Table 5-13
Example 4: Summary, Average Quantities (Class C Slurry)

TABULATION OF CALCULATED AVERAGE QUANTITIES:
Area weighted means of absolute values taken over
BOTTOM HALF of annular cross-section ...
O Average annular velocity = .6993E+02 in/sec
O Average apparent viscosity = .1812E-04 lbf sec/sq in
O Average stress, AppVis x dU/dx, = .1206E-02 psi
O Average stress, AppVis x dU/dy, = .1057E-02 psi
O Average dissipation = .4154E+00 lbf/(sec sq in)
O Average shear rate dU/dx = .1306E+03 1/sec
O Average shear rate dU/dy = .1110E+03 1/sec
O Average Stokes product = .1486E-02 lbf/in

TABULATION OF CALCULATED AVERAGE QUANTITIES, II:
Area weighted means of absolute values taken over
ENTIRE annular (x,y) cross-section ...
O Average annular velocity = .7129E+02 in/sec
O Average apparent viscosity = .1781E-04 lbf sec/sq in
O Average stress, AppVis x dU/dx, = .1076E-02 psi
O Average stress, AppVis x dU/dy, = .1078E-02 psi
O Average dissipation = .4097E+00 lbf/(sec sq in)
O Average shear rate dU/dx = .1181E+03 1/sec
O Average shear rate dU/dy = .1177E+03 1/sec
O Average Stokes product = .1494E-02 lbf/in

Example 5:   Concentric Rotating Flows, Stationary Baseline

While the cement slurry displaces drilling fluid in the annulus, the casing is often rotated from 10 - 20 rpm. The reasons are twofold: first to break the gel strength of any coagulated mud, and second to prevent flowing mud from gelling. The latter is amenable to flow simulation and modeling. It is of interest to compare the stress states for stationary versus rotating casings. We will assume a concentric geometry and use the analytical results of Chapter 3 obtained for rotating pipes. Again, the closed form expressions require a narrow annulus. The input summary shown in Table 5-14, extracted from output files, establishes reference conditions for a nonrotating flow run with "0.001 rpm". For stationary casings, the exact solution is of course available; however, for comparative reasons, we will not make use of it.

Table 5-14
Summary of Input Parameters

O Drill pipe outer radius (inches) =   4.0000
O Borehole radius (inches) =   5.0000
O Axial pressure gradient (psi/ft) =   .01000
O Drillstring rotation rate (rpm) =   .0010
O Drillstring rotation rate (rad/sec) =   .0001
O Fluid exponent "n" (nondimensional) =   .7240
O Consistency factor (lbf $\sec^n$/sq in) =   .1861E-04
O Mass density of fluid (lbf $\sec^2$/ft^4) =   1.9000
    (e.g., about 1.9 for water)
O Number of radial "grid" positions =   17

The software model calculates all physical quantities of possible interest. These include the variables discussed in Chapter 3, and in addition, several derivative flow properties. The complete roster of output variables is given in Table 5-15; this same chart is supplied with all computed results.

Table 5-15

Analytical (Non-Iterative) Solutions Tabulated vs "r"

Nomenclature and Units

---

r	Annular radial position	(in)
V_z	Velocity in axial z direction	(in/sec)
V_ϑ	Circumferential velocity	(in/sec)
$d\vartheta/dt$ or W	ϑ velocity	(rad/sec)
	(Note: 1 rad/sec = 9.5493 rpm)	
dV_z/dr	Velocity gradient	(1/sec)
dV_ϑ/dr	Velocity gradient	(1/sec)
dW/dr	Angular speed gradient	(1/(sec x in))
$S_{r\vartheta}$	$r\vartheta$ stress component	(psi)
S_{rz}	rz stress component	(psi)
S_{max}	Sqrt (S_{rz}**2 + $S_{r\vartheta}$**2)	(psi)
dP/dr	Radial pressure gradient	(psi/in)
	Note: 1 cp = 1.465E-7 lbf sec/sq in	
App-Vis	Apparent viscosity	(lbf sec/sq in)
Dissip	Dissipation function	(lbf/(sec x sq in))
	(indicates frictional heat produced)	
Atan V_ϑ/V_z ..	Angle between V_ϑ and V_z vectors	(deg)
Net Spd	Sqrt (V_z**2 + V_ϑ**2)	(in/sec)
$D_{r\vartheta}$	$r\vartheta$ deformation tensor component.	(1/sec)
D_{rz}	rz deformation tensor component..	(1/sec)

Calculations for the foregoing quantities at each of the 17 control points selected in Table 5-14 require only seconds on PC-XTs equipped with math co-processors. Various types of output are provided for correlation or research purposes. For example, "high level" results like

$$\text{Total volume flow rate (cubic in/sec)} = .2653E+03$$
$$\text{(gal/min)} = .6889E+02$$

Or "low level results" conveniently supplied in hybrid text and ASCII plot form for complete portability; for example, Figure 5-15a for annular velocity or Figure 5-15b for the "maximum stress" defined in Table 5-15.

```
      Axial speed Vz(r):
 r                   0
                     _____
5.00    -.2578E-05   I
4.94    .1383E+01    I*
4.88    .3393E+01    I   *
4.82    .5570E+01    I      *
4.76    .7731E+01    I         *
4.71    .9759E+01    I           *
4.65    .1156E+02    I             *
4.59    .1307E+02    I               *
4.53    .1421E+02    I                 *
4.47    .1494E+02    I                  *
4.41    .1519E+02    I                   *
4.35    .1493E+02    I                  *
4.29    .1410E+02    I                 *
4.24    .1267E+02    I               *
4.18    .1058E+02    I            *
4.12    .7797E+01    I         *
4.06    .4285E+01    I     *
4.00    .0000E+00    I
```

Figure 5-15a. Annular Velocity Distribution.

```
      Maximum stress Smax (r):
 r                   0
                     _____
5.00    .3958E-03    I                    *
4.94    .3512E-03    I                   *
4.88    .3062E-03    I                 *
4.82    .2606E-03    I               *
4.76    .2145E-03    I             *
4.71    .1678E-03    I           *
4.65    .1206E-03    I         *
4.59    .7282E-04    I *
4.53    .2443E-04    *
4.47    .2459E-04    *
4.41    .7426E-04    I *
4.35    .1246E-03    I    *
4.29    .1757E-03    I      *
4.24    .2275E-03    I        *
4.18    .2801E-03    I           *
4.12    .3334E-03    I             *
4.06    .3876E-03    I              *
4.00    .4427E-03    I                *
```

Figure 5-15b. Maximum Stress Distribution.

Detailed listings are also available for permanent reference; flow properties, tabulated against "r", can be read by spreadsheet programs for further trend analysis. A complete summary of calculated results appears in Tables 5-16 to 5-19.

Table 5-16
Averaged Values of Annular Quantities

Average Vz (in/sec) =   .9480E+01
         (ft/min) =   .4740E+02
Average Vθ (in/sec) = .3253E-02
Average W (rad/sec) = .7401E-03
Average total speed (in/sec) = .9480E+01
Average angle between Vz and Vθ (deg) = .2633E+01
Average d(Vz)/dr (1/sec) = .0000E+00
Average d(Vθ)/dr (1/sec) = -.8717E-02
Average dW/dr (1/(sec X in)) = -.2136E-02
Average dP/dr (psi/in) = .2709E-09
Average Srθ (psi) = .7355E-07
Average Srz (psi) = .7828E-05
Average Smax (psi) = .2097E-03
Average dissipation function (lbf sec^{n-2}/sq in) = .9225E-02
Average apparent viscosity (lbf sec^n/sq in) = .8692E-05
Average Drθ (1/sec) = .4729E-02
Average Drz (1/sec) = .8465E+00

Table 5-17
Detailed Tabulated Quantities

r	Vz	Vϑ	W	d(Vz)/dr	d(Vϑ)/dr	dW/dr
5.00	-.258E-05	-.967E-08	-.193E-08	-.682E+02	-.101E-01	-.203E-02
4.94	.138E+01	.587E-03	.119E-03	-.578E+02	-.980E-02	-.201E-02
4.88	.339E+01	.115E-02	.236E-03	-.478E+02	-.941E-02	-.198E-02
4.82	.557E+01	.170E-02	.352E-03	-.383E+02	-.894E-02	-.193E-02
4.76	.773E+01	.221E-02	.465E-03	-.293E+02	-.838E-02	-.186E-02
4.71	.976E+01	.270E-02	.574E-03	-.209E+02	-.768E-02	-.175E-02
4.65	.116E+02	.316E-02	.680E-03	-.132E+02	-.678E-02	-.161E-02
4.59	.131E+02	.358E-02	.780E-03	-.658E+01	-.554E-02	-.138E-02
4.53	.142E+02	.395E-02	.873E-03	-.146E+01	-.340E-02	-.944E-03
4.47	.149E+02	.428E-02	.958E-03	.147E+01	-.344E-02	-.984E-03
4.41	.152E+02	.455E-02	.103E-02	.676E+01	-.585E-02	-.156E-02
4.35	.149E+02	.475E-02	.109E-02	.138E+02	-.752E-02	-.198E-02
4.29	.141E+02	.487E-02	.113E-02	.222E+02	-.895E-02	-.235E-02
4.24	.127E+02	.487E-02	.115E-02	.317E+02	-.103E-01	-.270E-02
4.18	.106E+02	.473E-02	.113E-02	.423E+02	-.116E-01	-.305E-02
4.12	.780E+01	.437E-02	.106E-02	.538E+02	-.129E-01	-.340E-02
4.06	.428E+01	.360E-02	.888E-03	.663E+02	-.144E-01	-.376E-02
4.00	.000E+00	.419E-03	.105E-03	.796E+02	-.164E-01	-.413E-02

Table 5-18
Detailed Tabulated Quantities

r	Srϑ	Srz	Smax	dP/dr	App-Vis	Dissip
5.00	.588E-07	-.396E-03	.396E-03	.171E-20	.580E-05	.270E-01
4.94	.602E-07	-.351E-03	.351E-03	.639E-11	.607E-05	.203E-01
4.88	.617E-07	-.306E-03	.306E-03	.250E-10	.640E-05	.146E-01
4.82	.632E-07	-.261E-03	.261E-03	.547E-10	.680E-05	.998E-02
4.76	.648E-07	-.214E-03	.214E-03	.943E-10	.733E-05	.628E-02
4.71	.664E-07	-.168E-03	.168E-03	.142E-09	.805E-05	.350E-02
4.65	.681E-07	-.121E-03	.121E-03	.197E-09	.913E-05	.159E-02
4.59	.699E-07	-.728E-04	.728E-04	.256E-09	.111E-04	.479E-03
4.53	.717E-07	-.244E-04	.244E-04	.316E-09	.168E-04	.356E-04
4.47	.736E-07	.246E-04	.246E-04	.376E-09	.167E-04	.361E-04
4.41	.756E-07	.743E-04	.743E-04	.430E-09	.110E-04	.502E-03
4.35	.776E-07	.125E-03	.125E-03	.475E-09	.901E-05	.172E-02
4.29	.798E-07	.176E-03	.176E-03	.506E-09	.791E-05	.390E-02
4.24	.820E-07	.227E-03	.227E-03	.514E-09	.717E-05	.722E-02
4.18	.843E-07	.280E-03	.280E-03	.492E-09	.662E-05	.118E-01
4.12	.868E-07	.333E-03	.333E-03	.426E-09	.619E-05	.179E-01
4.06	.893E-07	.388E-03	.388E-03	.293E-09	.585E-05	.257E-01
4.00	.919E-07	.443E-03	.443E-03	.402E-11	.556E-05	.353E-01

Table 5-19
Detailed Tabulated Quantities

r	Vz	Vϑ	Atan Vϑ/Vz	NetSpd	Drϑ	Drz
5.00	-.258E-05	-.967E-08	.215E+00	.258E-05	.507E-02	-.341E+02
4.94	.138E+01	.587E-03	.243E-01	.138E+01	.496E-02	-.289E+02
4.88	.339E+01	.115E-02	.195E-01	.339E+01	.482E-02	-.239E+02
4.82	.557E+01	.170E-02	.175E-01	.557E+01	.465E-02	-.191E+02
4.76	.773E+01	.221E-02	.164E-01	.773E+01	.442E-02	-.146E+02
4.71	.976E+01	.270E-02	.159E-01	.976E+01	.413E-02	-.104E+02
4.65	.116E+02	.316E-02	.157E-01	.116E+02	.373E-02	-.661E+01
4.59	.131E+02	.358E-02	.157E-01	.131E+02	.316E-02	-.329E+01
4.53	.142E+02	.395E-02	.159E-01	.142E+02	.214E-02	-.728E+00
4.47	.149E+02	.428E-02	.164E-01	.149E+02	.220E-02	.735E+00
4.41	.152E+02	.455E-02	.172E-01	.152E+02	.344E-02	.338E+01
4.35	.149E+02	.475E-02	.182E-01	.149E+02	.431E-02	.691E+01
4.29	.141E+02	.487E-02	.198E-01	.141E+02	.504E-02	.111E+02
4.24	.127E+02	.487E-02	.221E-01	.127E+02	.572E-02	.159E+02
4.18	.106E+02	.473E-02	.256E-01	.106E+02	.637E-02	.212E+02
4.12	.780E+01	.437E-02	.321E-01	.780E+01	.700E-02	.269E+02
4.06	.428E+01	.360E-02	.482E-01	.428E+01	.763E-02	.331E+02
4.00	.000E+00	.419E-03	.886E+02	.419E-03	.827E-02	.398E+02

Example 6:   Concentric Rotating Flows, Rotating Casing

How do the computed results for non-rotating casing change if the casing were rotated at 20 rpm? Similar calculations were undertaken, with the following self-explanatory results in Tables 5-20 to 5-24.

Table 5-20
Summary of Input Parameters

O Drill pipe outer radius (inches) =   4.0000
O Borehole radius (inches) =   5.0000
O Axial pressure gradient (psi/ft) =   .01000
O Drillstring rotation rate (rpm) =   20.0000
O Drillstring rotation rate (rad/sec) =   2.0944
O Fluid exponent "n" (nondimensional) =   .7240
O Consistency factor (lbf sec^n/sq in) =   .1861E-04
O Mass density of fluid (lbf sec^2/ft^4) = 1.9000
  (e.g., about 1.9 for water)
O Number of radial "grid" positions =   17

Table 5-21
Averaged Values of Annular Quantities

Average Vz (in/sec)   =   .1008E+02
      (ft/min)   =   .5040E+02
Average Vϑ (in/sec) =   .4987E+01
Average W (rad/sec)   =   .1146E+01
Average total speed (in/sec) = .1157E+02
Average angle between Vz and Vϑ (deg) = .2711E+02
Average d(Vz)/dr (1/sec) = .0000E+00
Average d(Vϑ)/dr (1/sec) = -.1217E+02
Average dW/dr (1/(sec X in)) = -.3005E+01
Average dP/dr (psi/in) = .6528E-03
Average Srϑ (psi) = .9562E-04
Average Srz (psi) = .7828E-05
Average Smax (psi) = .2386E-03
Average dissipation function (lbf sec^{n-2}/sq in) = .1073E-01
Average apparent viscosity (lbf sec^n/sq in) = .7458E-05
Average Drϑ (1/sec) = .6656E+01
Average Drz (1/sec) = .9358E+00

Observe that our calculated "maximum stresses" S_{max} have increased significantly from those of Example 5 without rotation. This increase may thwart flowing mud from gelling; that is, prevent the cement slurry from channeling and bypassing isolated pockets of resistive gelled mud. The closed form solutions provide a means to quickly estimate changes to flowing properties, and are valuable in this respect. Again, the model predicts total volume flow rate. In the present case, the result

$$\text{Total volume flow rate (cubic in/sec)} = .2828E+03$$
$$(\text{gal/min}) = .7345E+02$$

exceeds that obtained for non-rotating casing of Example 5 under the same pressure drop. This is physically consistent with well known effects of rotation. Detailed plots are available for any of the tabulated quantities, e.g., Figures 5-16a and 5-16b for velocity and stress.

Table 5-22
Detailed Tabulated Quantities

r	Vz	Vϑ	W	d(Vz)/dr	d(Vϑ)/dr	dW/dr
5.00	-.397E-04	-.127E-04	-.253E-05	-.687E+02	-.133E+02	-.266E+01
4.94	.235E+01	.769E+00	.156E+00	-.584E+02	-.129E+02	-.263E+01
4.88	.459E+01	.151E+01	.310E+00	-.485E+02	-.124E+02	-.260E+01
4.82	.675E+01	.223E+01	.462E+00	-.390E+02	-.118E+02	-.255E+01
4.76	.883E+01	.292E+01	.612E+00	-.301E+02	-.112E+02	-.248E+01
4.71	.107E+02	.357E+01	.758E+00	-.218E+02	-.105E+02	-.238E+01
4.65	.124E+02	.418E+01	.900E+00	-.143E+02	-.963E+01	-.227E+01
4.59	.138E+02	.476E+01	.104E+01	-.787E+01	-.878E+01	-.214E+01
4.53	.149E+02	.528E+01	.117E+01	-.246E+01	-.821E+01	-.207E+01
4.47	.155E+02	.576E+01	.129E+01	.250E+01	-.843E+01	-.217E+01
4.41	.157E+02	.619E+01	.140E+01	.820E+01	-.945E+01	-.246E+01
4.35	.153E+02	.656E+01	.151E+01	.152E+02	-.108E+02	-.283E+01
4.29	.144E+02	.687E+01	.160E+01	.235E+02	-.123E+02	-.323E+01
4.24	.129E+02	.713E+01	.168E+01	.330E+02	-.138E+02	-.365E+01
4.18	.108E+02	.736E+01	.176E+01	.435E+02	-.153E+02	-.407E+01
4.12	.793E+01	.760E+01	.184E+01	.550E+02	-.167E+02	-.451E+01
4.06	.435E+01	.791E+01	.195E+01	.674E+02	-.182E+02	-.497E+01
4.00	.000E+00	.838E+01	.209E+01	.807E+02	-.197E+02	-.545E+01

Table 5-23
Detailed Tabulated Quantities

r	Srϑ	Srz	Smax	dP/dr	App-Vis	Dissip
5.00	.765E-04	-.396E-03	.403E-03	.294E-14	.576E-05	.282E-01
4.94	.783E-04	-.351E-03	.360E-03	.110E-04	.602E-05	.215E-01
4.88	.802E-04	-.306E-03	.316E-03	.430E-04	.632E-05	.159E-01
4.82	.822E-04	-.261E-03	.273E-03	.944E-04	.668E-05	.112E-01
4.76	.842E-04	-.214E-03	.230E-03	.163E-03	.713E-05	.744E-02
4.71	.863E-04	-.168E-03	.189E-03	.248E-03	.769E-05	.463E-02
4.65	.885E-04	-.121E-03	.150E-03	.345E-03	.841E-05	.266E-02
4.59	.908E-04	-.728E-04	.116E-03	.452E-03	.925E-05	.147E-02
4.53	.932E-04	-.244E-04	.964E-04	.565E-03	.994E-05	.934E-03
4.47	.957E-04	.246E-04	.988E-04	.681E-03	.985E-05	.991E-03
4.41	.982E-04	.743E-04	.123E-03	.795E-03	.905E-05	.168E-02
4.35	.101E-03	.125E-03	.160E-03	.905E-03	.819E-05	.314E-02
4.29	.104E-03	.176E-03	.204E-03	.101E-02	.747E-05	.557E-02
4.24	.107E-03	.227E-03	.251E-03	.110E-02	.690E-05	.915E-02
4.18	.110E-03	.280E-03	.301E-03	.119E-02	.644E-05	.140E-01
4.12	.113E-03	.333E-03	.352E-03	.128E-02	.607E-05	.204E-01
4.06	.116E-03	.388E-03	.405E-03	.141E-02	.575E-05	.285E-01
4.00	.120E-03	.443E-03	.459E-03	.161E-02	.549E-05	.383E-01

Table 5-24
Detailed Tabulated Quantities

r	Vz	Vϑ	AtanVϑ/Vz	NetSpd	Drϑ	Drz
5.00	-.397E-04	-.127E-04	.177E+02	.416E-04	.664E+01	-.344E+02
4.94	.235E+01	.769E+00	.181E+02	.247E+01	.651E+01	-.292E+02
4.88	.459E+01	.151E+01	.183E+02	.483E+01	.635E+01	-.242E+02
4.82	.675E+01	.223E+01	.183E+02	.711E+01	.615E+01	-.195E+02
4.76	.883E+01	.292E+01	.183E+02	.930E+01	.591E+01	-.150E+02
4.71	.107E+02	.357E+01	.184E+02	.113E+02	.561E+01	-.109E+02
4.65	.124E+02	.418E+01	.186E+02	.131E+02	.527E+01	-.717E+01
4.59	.138E+02	.476E+01	.190E+02	.146E+02	.491E+01	-.394E+01
4.53	.149E+02	.528E+01	.195E+02	.158E+02	.469E+01	-.123E+01
4.47	.155E+02	.576E+01	.204E+02	.166E+02	.486E+01	.125E+01
4.41	.157E+02	.619E+01	.215E+02	.169E+02	.543E+01	.410E+01
4.35	.153E+02	.656E+01	.231E+02	.167E+02	.616E+01	.761E+01
4.29	.144E+02	.687E+01	.254E+02	.160E+02	.694E+01	.118E+02
4.24	.129E+02	.713E+01	.289E+02	.148E+02	.772E+01	.165E+02
4.18	.108E+02	.736E+01	.343E+02	.130E+02	.851E+01	.217E+02
4.12	.793E+01	.760E+01	.438E+02	.110E+02	.929E+01	.275E+02
4.06	.435E+01	.791E+01	.612E+02	.902E+01	.101E+02	.337E+02
4.00	.000E+00	.838E+01	.900E+02	.838E+01	.109E+02	.404E+02

Axial speed Vz(r):

r		0
5.00	-.3966E-04	I
4.94	.2350E+01	I *
4.88	.4586E+01	I *
4.82	.6754E+01	I *
4.76	.8827E+01	I *
4.71	.1074E+02	I *
4.65	.1243E+02	I *
4.59	.1383E+02	I *
4.53	.1488E+02	I *
4.47	.1551E+02	I *
4.41	.1568E+02	I *
4.35	.1534E+02	I *
4.29	.1444E+02	I *
4.24	.1293E+02	I *
4.18	.1077E+02	I *
4.12	.7926E+01	I *
4.06	.4348E+01	I *
4.00	.0000E+00	I

Figure 5-16a. Annular Velocity Distribution.

Maximum stress Smax (r):

r		0
5.00	.4032E-03	I *
4.94	.3599E-03	I *
4.88	.3165E-03	I *
4.82	.2732E-03	I *
4.76	.2304E-03	I *
4.71	.1887E-03	I *
4.65	.1496E-03	I *
4.59	.1164E-03	I *
4.53	.9636E-04	I *
4.47	.9879E-04	I *
4.41	.1232E-03	I *
4.35	.1604E-03	I *
4.29	.2040E-03	I *
4.24	.2512E-03	I *
4.18	.3007E-03	I *
4.12	.3520E-03	I *
4.06	.4046E-03	I *
4.00	.4586E-03	I *

Figure 5-16b. Maximum Stress Distibution.

COILED TUBING RETURN FLOWS

After a well has been drilled and cemented, and after it has seen production, fines and sands may have emerged through the perforations. This may be the case with unconsolidated sands and unstable wellbores; in deviated and horizontal wells, the debris remains on the low side of the hole. To remove these fines, metal tubing unrolled from "coils" at the surface (typically, 1-in. to 2-in. O.D.) is run to the required depth. Fluids are pumped downhole through this tubing. They clean the highly eccentric annulus, and return to the surface carrying the debris.

The clean-up process is not unlike cuttings transport in drilling, except that hole eccentricities here are more severe. Besides sand washing, coiled tubing is also used in paraffin cleanout, acid or cement squeezes, and mud displacement. Typical fluids may include nitrogen or non-Newtonian foams. Figure 5-17a displays a typical annulus encountered in coiled tubing applications; note the large diameter ratio and the typically high eccentricities.

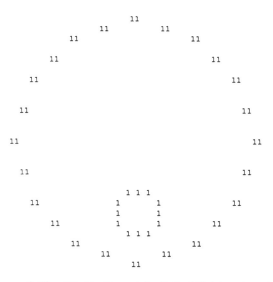

Figure 5-17a. Highly Eccentric Coiled Tubing Annulus.

The eccentric flow model of Chapter 2 can be used to simulate fluid motions in such annuli. Figures 5-17b and 5-17c, for example, display the boundary conforming mesh generated for this system and a typical velocity field computed in these coordinates. As in cuttings transport, velocity and stress are expected to play important roles in sand cleaning. Annular flow simulation, again, is straightforward and robust; the eccentric model calculates the required quantities accurately. Note how no-slip velocity boundary conditions are exactly satisfied at all solid surfaces.

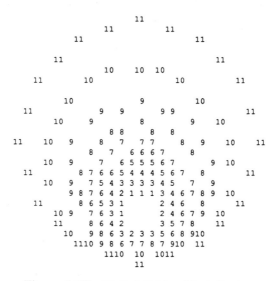

Figure 5-17b. Coiled Tubing Mesh System.

HEAVILY CLOGGED STUCK PIPE

So far we have considered *annular* flows only, that is, dough-nut shaped "doubly-connected" geometries. How can we model "singly connected" *pipe-like* geometries, not atypical of annuli highly clogged by thick cuttings beds, without completely reformulating the problem? Figure 5-18a shows one possible clogged configuration, where we have displayed the *drillpipe plus cuttings bed* as a *single* entity whose boundary is marked by asterisks.

At the lower annulus, the separation from the borehole bottom is kept to a token minimum, say 0.01 inch. The computer program will give nearly zero velocities for this narrow gap since no-slip conditions predominate. For all practical purposes, the gap is impermeable to flow and completely plugs up the bottom.

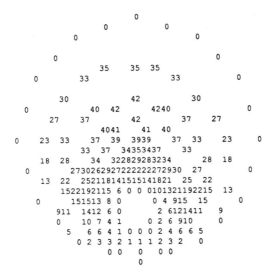

Figure 5-17c. Annular Velocity.

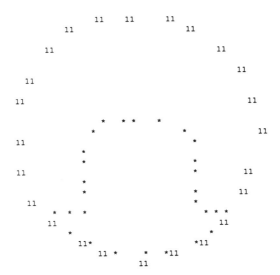

Figure 5-18a. Heavily Clogged Annulus.

Hence, the only flow domain of dynamical significance is the simple region just above the thick cuttings bed. The corresponding mesh system and and a typical velocity field are shown in Figures 5-18b and 5-18c for a power law mud.

```
              11      11      11
          11          10          11
                  10      10
      11      10      9   9   9   10    11
          10    9       8   8   8     9  10
                  8   7   7     7  8           11
      11      9       7  6   6     6   7     9
        10      8   6     5 5   5     6   8    10
      11      9   7 6   5   4 4   4   5   6 7   9    11
        10    8 7   5   4   3 2   3   4   5   7 8  10
          9    6   4 3 2   1 1   1 1 2 3 4     6    9
          8 7   5   2               2   5   7 8      11
      11      6 5   3                   3   5 6     10
        9 8     4 3 1                   3 4     8 9
            7 5 2 1                     1 2 5 7
    1110      4 3                       3 4     1011
        8 7 5   1                       5 7 8
            4 3 1                       1 3 4     11
    1110 8 5 3                          1 3 5 810
          1     1                       1 1
        11 1                                11
                                          1
          11                           1011
              11 1      10      11
                        11
```

Figure 5-18b. Computed Mesh System.

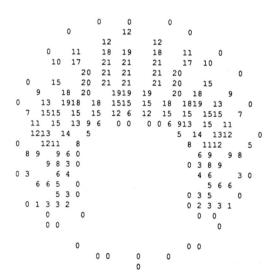

Figure 5-18c. Annular Velocity.

CONCLUSIONS

We have approached the subject of annular flow quite generally in Chapters 2, 3 and 4, and developed models for different geometries in several physical limits for Newtonian and non-Newtonian rheologies. These models were used in the present chapter to study several drilling and production problems of importance in deviated and horizontal wells. These included cuttings transport, spotting fluid analysis, cementing and coiled tubing applications. In each case, our limited studies addressed only the annular flow simulation aspects of the problem. We claim only that the models developed here are useful in correlating real world behaviour with flow properties that, until now, have resisted analysis; and that our studies appear to have identified some potentially useful ideas.

With respect to our eccentric flow solutions, we showed here and in Chapter 2 that the numerical model can be run "as is" and still *generate physically correct qualitative trends and quantitative estimates.* Mesh dependence, of importance at least academically, is not significant in this respect. This robustness is very important from a practical and operational viewpoint, since few exact solutions are ever available for mesh calibration.

We emphasize that the use of the rotational viscometer, or any other mechanical viscometer, should be restricted to determining the *intrinsic* rheological parameters (such as "n" or "k") that characterize a particular non-Newtonian fluid. Well-designed viscometers are ideal for this purpose. They are, however, not suitable for predicting actual downhole properties during specific drilling and cementing runs because actual shear rates, which can and do vary widely, are always unknown a priori. This explains why conventional attempts to correlate cuttings transport data with viscometer properties have been unsuccessful.

A viscometer that actually predicts downhole properties is, of course, difficult to construct. Given all the nonlinearities and nuances of non-Newtonian flow and borehole eccentricity, it is unduly optimistic to believe that a dimensionally scalable mechanical device can be built at all. But that is not necessary. As we have amply demonstrated, software simulation provides an elegant and efficient means to achieve most drilling and production objectives. And it can be improved upon as new techniques and insights materialize.

We have dealt only with the drilling and production aspects of annular flow. These, of course, represent small subsets of industrial application. Annular flow is also important in chemical engineering, manufacturing and extrusion processes. It is clear that the methods developed in this book can be applied to disciplines outside the petroleum industry, and an effort is underway to explore these possibilities.

REFERENCES

---

1. Chin, W.C., "Advances in Annular Borehole Flow Modeling," *Offshore Magazine*, February 1990, pp. 31-37.

2. Becker, T.E., Azar, J.J., and Okrajni, S.S., "Correlations of Mud Rheological Properties with Cuttings Transport Performance in Directional Drilling," *SPE Paper 19535, 64th Annual Technical Conference and Exhibition, Society of Petroleum Engineers*, San Antonio, October 1989.

3. Chin, W.C., "Exact Cuttings Transport Correlations Developed for High Angle Wells," *Offshore Magazine*, May 1990, pp. 67-70.

4. Streeter, V.L., *Handbook of Fluid Dynamics*, New York: McGraw-Hill, 1961.

5. Seeberger, M.H., Matlock, R.W., and Hanson, P.M., "Oil Muds in Large Diameter, Highly Deviated Wells: Solving the Cuttings Removal Problem," *SPE/IADC Paper 18635, 1989 SPE/IADC Drilling Conference, New Orleans*, February 28 - March 3, 1989.

6. Fraser, L.J., "Field Application of the All-Oil Drilling Fluid Concept," *IADC/SPE Paper 19955, 1990 IADC/SPE Drilling Conference*, Houston, February 27 - March 2, 1990.

7. Fraser, L.J., "Green Canyon Drilling Benefits from All Oil Mud," *Oil and Gas Journal*, March 19, 1990, pp. 33-39.

8. Fraser, L.J., "Effective Ways to Clean and Stabilize High-Angle Hole," *Petroleum Engineer International*, November 1990, pp. 30-35.

9. Chin, W.C., "Annular Flow Model Explains Conoco's Borehole Cleaning Success," *Offshore Magazine*, October 1990, pp. 41-42.

10. Quigley, M.S., Dzialowski, A.K., and Zamora, M., "A Full-Scale Wellbore Friction Simulator," *IADC/SPE Paper 19958, 1990 IADC/SPE Drilling Conference, Houston*, February 27 - March 2, 1990.

11. Brown, N.P., Bern, P.A., and Weaver, A., "Cleaning Deviated Holes: New Experimental and Theoretical Studies," *SPE/IADC Paper 18636, 1989 SPE/IADC Drilling Conference, New Orleans*, February 28 - March 3, 1989.

12. Gray, K.E., "The Cutting Carrying Capacity of Air at Pressures Above Atmosphere," *Petroleum Transactions, AIME, Vol. 213*, 1958, pp. 180-185.

13. Adewumi, M.A., and Tian, S., "Hydrodynamic Modeling of Wellbore Hydraulics in Air Drilling," *SPE Paper 19333, 1989 SPE Eastern Regional Meeting, Morgantown*, October 24-27, 1989.

14. Martin, M., Georges, C., Bisson, P., and Konirsch, O., "Transport of Cuttings in Directional Wells," *SPE/IADC Paper 16083, 1987 SPE/IADC Drilling Conference, New Orleans,* March 15-18, 1987.

15. Tomren, P.H., Iyoho, A.W., and Azar, J.J., "Experimental Study of Cuttings Transport in Directional Wells," *SPE Drilling Engineering,* February 1986, pp. 43-56.

16. Okragni, S.S., "Mud Cuttings Transport in Directional Well Drilling," *SPE Paper 14178, 60th Annual Technical Conference and Exhibition of the Society of Petroleum Engineers, Las Vegas,* September 22-25, 1985.

17. Hussaini, S.M., and Azar, J.J., "Experimental Study of Drilled Cuttings Transport Using Common Drilling Muds," *Society of Petroleum Engineers Journal,* February 1983, pp. 11-20.

18. Ford, J.T., Peden, J.M., Oyeneyin, M.B., Gao, E., and Zarrough, R., "Experimental Investigation of Drilled Cuttings Transport in Inclined Boreholes," *SPE Paper 20421, 65th Annual Technical Conference and Exhibition of the Society of Petroleum Engineers, New Orleans,* September 23-26, 1990.

19. Sifferman, T.R., and Becker, T.E., "Hole Cleaning in Full-Scale Inclined Wellbores," *SPE Paper 20422, 65th Annual Technical Conference and Exhibition of the Society of Petroleum Engineers, New Orleans,* September 23-26, 1990.

20. Harvey, F., "Fluid Program Built Around Hole Cleaning Protecting Formation," *Oil and Gas Journal,* 5 November 1990, pp. 37-41.

21. Seheult, M., Grebe, L., Traweek, J.E., and Dudley, M., "Biopolymer Fluids Eliminate Horizontal Well Problems," *World Oil,* January 1990, pp. 49-53.

22. Savins, J.G., and Wallick, G.C., "Viscosity Profiles, Discharge Rates, Pressures, and Torques for a Rheologically Complex Fluid in a Helical Flow," *A.I.Ch.E. Journal, Vol. 12, No. 2,* March 1966, pp. 357-363.

23. Outmans, H.D., "Mechanics of Differential Pressure Sticking of Drill Collars," *Petroleum Transactions, AIME, Vol. 213,* 1958, pp. 265-274.

24. Halliday, W.S., and Clapper, D.K., "Toxicity and Performance Testing of Non-Oil Spotting Fluid for Differentially Stuck Pipe," *Paper 18684, SPE/IADC Drilling Conference, New Orleans,* February 28 - March 3, 1989.

25. Chin, W.C., "Model Offers Insight into Spotting Fluid Performance," *Offshore Magazine,* February 1991, pp. 32-33.

26. Smith, D.K.,*Cementing,* Dallas: Society of Petroleum Engineers, 1976.

27. Suman, G.O., and Ellis, R.C., *World Oil's Cementing Handbook,* Houston: Gulf Publishing Company, 1977.

28. Suman, G.O., *Cementing,* Houston: Completion Tool Company, 1990.

29. Baret, F., Free, D.L., and Griffin, T.J., "Hard and Fast Rules for Effective Cementing," *Drilling Magazine,* March-April 1989, pp. 22-25.

30. Zaleski, T.E., and Ashton, J.P., "Gravel Packing Feasible in Horizontal Well Completions," *Oil and Gas Journal*, 11 June 1990, pp. 33-37.

31. Smith, T.R., "Cementing Displacement Practices - Field Applications," *Journal of Petroleum Technology*, May 1990, pp. 564-566.

32. Wilson, M.A., and Sabins, F.L., "A Laboratory Investigation of Cementing Horizontal Wells," *SPE Drilling Engineering*, September 1988, pp. 275-280.

33. Zaleski, T.E., and Spatz, E., "Horizontal Completions Challenge for Industry," *Oil and Gas Journal*, 2 May 1988, pp. 58-70.

34. Suman, G.O., and Snyder, R.E., "Primary Cementing: Why Many Conventional Jobs Fail," *World Oil*, December 1982.

35. Lockyear, C.F., Ryan, D.F., and Gunningham, M.M., "Cement Channeling: How to Predict and Prevent," *SPE Drilling Engineering*, September 1990, pp. 201-208.

36. George, C., "Innovations Change Cementing Operations," *Petroleum Engineering International*, October 1990, pp. 37- 41.

37. Benge, G., "Field Study of Offshore Cement-Spacer Mixing," *SPE Drilling Engineering*, September 1990, pp. 196-200.

38. Sparlin, D.D., and Hagen, R.W., "Controlling Sand in a Horizontal Completion," *World Oil*, November 1988, pp. 54-58.

39. Suman, G. O., "Cementing - The Drilling/Completion Interface," Completion Tool Company Technical Report, September 1988.

40. Lin, C.C., *The Theory of Hydrodynamic Stability*, London: Cambridge University Press, 1967.

Index